THE UNITY OF SCIENCE

The Unity of Science

Exploring Our Universe,
from the Big Bang to the
Twenty-First Century

IRWIN SHAPIRO

Yale
UNIVERSITY PRESS
NEW HAVEN AND LONDON

Published with assistance from the foundation established in memory
of Amasa Stone Mather of the Class of 1907, Yale College.

Yale University Press books may be purchased in quantity for educational, business,
or promotional use. For information, please e-mail sales.press@yale.edu (U.S. office)
or sales@yaleup.co.uk (U.K. office).

Set in Electra type by Newgen North America.
Printed in the United States of America.

Library of Congress Control Number: 2023931521
ISBN 978-0-300-25361-0 (hardcover : alk. paper)

A catalogue record for this book is available from the British Library.

This paper meets the requirements of ANSI/NISO Z39.48-1992
(Permanence of Paper).

10 9 8 7 6 5 4 3 2 1

CONTENTS

CONTENTS

CONTENTS

M ore than a decade ago, I created a one-semester course for Harvard
University undergraduates in the humanities and social sciences,
from first years to seniors. I had wanted to provide them with an overview
of science under the alliterative rubric, originally the title of this book: *The
Unity of Science: From the Big Bang to the Brontosaurus and Beyond.* My
goal, to consider why science is important in the context of our lives, fo-
cused on the importance of asking good questions.

Originally, I had no thought of turning the course into a book. My students, however, had other ideas and complained about the lack of one. After
a few years of such objections, I capitulated and started to write chapters,
posting each after its discussion in class. Then, out of the blue, so to say,
Joseph Calamia, then of Yale University Press, approached me about the
possibility of publishing the chapters as a book. One thing led to the next,
and here we are.

This work is an introduction for the curious reader who wants to learn
how, over the course of human history, science has advanced our understanding of the natural world. The chapters are organized thematically: the
first theme is concerned with unveiling the universe, or looking up; the
second with the earth and its fossils, or looking down; and the third with
the story of life, or looking in. Mathematics is used sparingly in the book.
Those readers not exactly fond of math nonetheless need not worry. One

can understand virtually everything in the book while completely ignoring the math.

I am not an expert in every subject that I discuss, not nearly so. I therefore asked experts to comment on the various aspects for which I felt insufficiently prepared and several generalists to read the book and provide their views. I gratefully acknowledge here, in alphabetical order, each of those very generous people who responded to my queries and made useful comments: Walter Alvarez on the demise of dinosaurs; Kay Behrensmeyer on the history of fossils; Roy Blount on the experiment in heredity; Steven Brusatte on characteristics of sauropods; Michael D'Emic on the lack of sauropod tooth fossils; Michael Foley on his careful reading of the entire book; Owen Gingerich on issues concerning the classical Greek contributions to our knowledge of solar system behavior; Philip Gingerich on the conversion of artiodactyls from land animals to sea mammals (whales); Bob Goldberg on the story of tardigrades; Alyssa Goodman on the formation of the earth; Megan Hochstrasser on her patient descriptions for me of the intricacies of CRISPR-Cas9; Emile Kreveld on details surrounding the discovery of the finite speed of light; Jonathan Losos on the experiment in heredity; Mike McCormick on the plague pandemic of the Justinian era; Mark Richards on the possible contributions of the Deccan Traps to the dinosaur demise; Doug Robertson on many topics that used the amazing breadth and depth of his scientific knowledge; and Bob Wilson on the cosmic microwave background radiation. Joshua Winn's editing of the manuscript was extraordinarily thorough and useful.

Maria McEachern edited the book with care, and my friend Bob Reasenberg and my family members—granddaughter Elina Shapiro, daughter Nancy Shapiro, son Steven Shapiro, and grandson Zev Shapiro—made extremely useful contributions. My grandson David Shapiro was enormously helpful in obtaining illustration permissions and for redrawing and improving most of them. Reviewers of the manuscript for Yale University Press, including Frank Shu, who disclosed his identity, made helpful and important points. Senior executive editor Jean Thomson Black, editorial assistant

Elizabeth Sylvia, and copyeditor Laura Jones Dooley played key roles in ensuring that this book made it into print.

I would be amazed, nonetheless, if there were no errors in the final product: my net was cast too far and wide not to have some at least tiny holes through which errors crept. I hope nonetheless that readers will find the book both interesting and informative and that their appreciation of science and the key role it plays, and will continue to play, in all of our lives is increased.

One point must be borne in mind: science advances continually if not continuously, so that new insights and discoveries will keep surfacing, outdating, and replacing some of the information already uncovered and contained in this book, which at best does not go beyond what is presently known. Since I am not noted for my clairvoyant powers, my predictions on new insights and discoveries are unlikely to be on target and so I have omitted making any specific projections.

I wish to express my undying thanks to my wife, Marian, who supported me throughout with wise suggestions and enormous amounts of her valuable time in editing, as well as to my children and grandchildren, who urged me on and did not complain at lost time with their father and grandfather, respectively.

Last, I again thank Maria McEachern, who, since I began to teach the course on which this book is based, has unflaggingly and always promptly given me incredible support in obtaining everything I needed from books and journals virtually worldwide. Without her help, I never could or would have written this book.

THE UNITY OF SCIENCE

Why Should We Care about
the Unity of Science?

M ost of us tend not to think about science in our everyday lives. Yet were we to stop to think, we'd realize soon enough that science is at the very heart of how we live. It underlies almost all of the devices that we are so utterly dependent upon. From the alarms that rouse us in the morning, through the electric toothbrush that massages our gums, through the toaster that warms our breakfast bagel, through the computer and the internet that brings our email, through the smartphone that has a semi-infinite number of applications on which we so heavily rely, to the now computer-assisted car that provides many of us with most of our daily transportation, and on and on . . . you get the point.

In this book, I intend to provide you with very broad science content in a coherent, semihistorical manner. I will discuss, among many other topics, the surprising presence in the world of dark matter and dark energy, the largest creatures to ever walk the earth, the sauropods, and the application of molecular biology to the Black Death. Whether my intentions are fulfilled is for you to decide. An underlying message, which I am selling, is the importance to the advancement of science of asking good, probing questions, fundamental to the extent feasible. Throughout, I will be posing questions; by no means all, though, will qualify as fundamental.

Before launching into the subject matter of the book, I will outline its philosophy, including my intended multiple meanings of "the unity of science." But first let's give "science" a definition, useful for this book. In essence, science is concerned with how nature behaves—that is, how the so-called natural world works—and how we can model this behavior. More specifically and somewhat simplistically, we can divide science into two parts: in the first, we observe, poke, and prod nature to determine how it behaves; and in the second, we use our minds to develop models that capture this behavior in a compact form and accurately predict not yet observed phenomena.

This seems like a sensible, clean separation: nature's behavior on one side and human-created models making sense of this behavior on the other. Is anything wrong with this picture? Yes, we spoil the works. Humans are in the middle and on both ends. Our brains, with the aid of our senses, are, in effect, the observers of nature's behavior and are also the creators of the models. We cannot extract ourselves from the loop. And we would dearly like to understand much better how our brains work!

A further subtlety: most contemporary observations and experiments are far from strictly determinations of nature's behavior. Rather, they are intricate mixtures of observations and deductions from our usually very well established models of nature's behavior. This mixture is a necessary aspect of all observations and/or experiments. After all, models are needed even to read the markings on a ruler. For some observations and experiments it is easier to understand this mix than for others. Think, for example, of the claim of discovery of a particle with the Large Hadron Collider in Switzerland and France—the collider extends under both countries! Only one property of this particle was determined at first: its mass. The value of the mass found was in the range expected for a Higgs boson (rhymes with "nose on"). Despite the incredible complexity and intertwining in this experiment between observations and models, no one questions (well, there are always some who make a curmudgeonly career out of questioning everything) that this discovery is of a real property of nature. Why? Partly because the models used in analyzing the experiment have been thoroughly established through past experience, and mostly because two independent groups, each with its own detection equipment, obtained the same result, within their estimated uncertainties.

These intrinsic measurement-model separation problems notwithstanding, I think it is nonetheless very helpful to view science in this bipartite manner. In other words, we divide science into two parts: evidence and inference. Evidence encompasses how nature behaves and inference involves our models for this behavior.

Thus we can say how nature works, but can we say why? That is, we can describe how nature behaves from observing it suitably. But can we say why nature works the way we observe it to work? We might have a model that adequately describes nature's behavior for a class of observed behaviors. But we can't say that this model is the reason why nature behaves in that way, because, for example, our model is not necessarily unique. Moreover, there is no known way to prove that a model is unique in being able to describe nature's behavior, and in that sense, we cannot address this type of "why?" question. There are, though, other types of "why?" questions that we can address. We can ask why a thrown ball will follow the particular path that it does follow, and we can provide a quite acceptable answer within the context of our model of projectile motions. But that is different from saying why nature behaves in that manner; it is, in effect, just another way of saying, or describing, that it does behave in that manner. Ultimately, we can establish only that our models "work" to the levels we have tested them (or don't work and need to be superseded by ones that work better).

The Unity of Science

What do we mean by the unity of science? Let us describe some of our meanings:

1. There is only one world (or universe, if you prefer the less colloquial term) that we observe. We have subdivided science into different disciplines the better to concentrate our efforts and make progress in understanding this world—that is, nature's behavior. But we should not lose sight that all parts of our world interact with each other at some level and are in that sense united.

2. The so-called scientific method is common to all of the disciplines of science in that we use it in each. This method can be considered as the way we use logical thinking to determine nature's behavior, by observation and/or experiment, and to develop models of this behavior that have predictive power (check, for example, the Web for sharper and more expansive definitions; the above is home-grown skimpy).

3. Phenomena that occur in one discipline, or domain, of science can have key effects on another, and often vice versa as well. For example, consider the changes that have occurred in the percentage of oxygen in our atmosphere. Biological processes are believed to be responsible for an apparently rather dramatic increase over two billion years ago in the atmosphere's oxygen content. This increase in turn affected the environment's influence on further biological evolution that led to us, and also perforce affected atmospheric science.

4. Multidisciplinary approaches are important to solving scientific problems. An example is the joint use of paleontology, ontology (= fetal development), and molecular biology in uncovering the history of the evolution of mammals, such as polar bears and whales.

5. Discoveries made, and tools developed, in one area of science enable advances to be made in other areas of science. We encounter many such examples in this book. Here is just one: radioactivity, developed by physicists and used by earth scientists to determine the age of the earth.

6. The diffusion of scientists trained in one area of science to work in another is common. For example, in the last half of the twentieth century, many physicists morphed into astronomers; in the twenty-first, many chemists, along with physicists and mathematicians/statisticians, are turning into molecular biologists.

These various meanings, especially the middle two, overlap to some extent. Nevertheless, they have somewhat different nuances, which are helpful to distinguish so as to enable us to have a broader and fuller understanding of the unity of science.

Models

Our models of nature's behavior are useful only insofar as they can accurately predict behaviors of nature not yet observed. If models could do no more than predict what has already been observed, they would not give us any additional "understanding" of nature that we could use to better harness it. Of course, we continually test each model against nature; when a model is found wanting, we seek a replacement with one of better utility. We discuss examples in later chapters.

Suppose two or more models predict phenomena equally well. How do we choose among them? Although this situation doesn't often, if at all, arise in practice, there's a well-discussed method to treat it. Most scientists and philosophers—no, I haven't taken a poll, it's just an impression based on my experience—subscribe to some form of Occam's razor: one chooses the simplest of the competing models. Of course, the definition of "simplest" may reside somewhat in the eye of the beholder; there's no universal agreement. We'll stop this discussion here, before we turn this book from science completely into cocktail-party philosophy.

Last, words about scientists' use of language in descriptions of models: first, there is no language police. The words scientists use to describe models often bear little relation to what you might naively think from the everyday usage of these words. Consider samples: principle, law, rule, theory, hypothesis, speculation, and guess. You might assume that these descriptors are listed in (approximate) order of their reliability, in the following sense: a model termed a law, which must be inviolable, down to a model termed a guess, which is far more likely to be wrong. This presumption is reasonable but hardly reliable. A now classic example that we shall encounter concerns a law about the spacing of our planets that is anything but. Still later we'll deal with a theory that accurately encompasses far more natural phenomena in its predictions than a related law. In other words, in this case, which deals with gravitation, a theory is a far better model of nature's behavior than is the corresponding law!

Examples of Nature's Behavior

Let's now delve slightly below these surfaces via two concrete examples: in the first, I hold a fork on my right thumb (fig. I.1). Suppose I pull my thumb away. What will happen? Most, if not all, adults will respond, "The fork will fall to the floor." Now I ask, "Why?" Most educated adults will reply, "Because of gravity." That is where they would fall—off my track. The fork falls to the floor *because that is the way nature behaves.* And this behavior occurs independently of whether we have created a model for it and regardless of the name we call it. It is a perhaps subtle but nonetheless fundamental distinction that we are making here and is the basic rationale for our largely eliminating the question "why?" from this book: "ultimate causes" lie, at least now, outside the realm of science. Such suggested causes are not subject to being tested—that is, are not "falsifiable"—and therefore are

Figure I.1. Irwin holding a fork, just before withdrawing his thumb from under the fork. Courtesy of the author.

not considered to be science by most scientists (not that I've taken a poll on this issue either . . .). There will, of course, always be a usually very small minority who will have a different opinion.

In the second, I hold a golf ball in one hand and a softball in the other, both at the same height from the ground. I let go of both at virtually the same instant. What happens? As with the fork, both fall. But miraculously, they strike the ground simultaneously—they seem to fall at the same rate. This, too, is the way nature behaves. We will see how this result appears, or is incorporated, in our models.

With these understandings of nature's behavior comfortably residing within our brains, we move on to a related subject.

The Consistency of Nature and the Characteristics of Science

It is common to think that nature's behavior is repeatable. I let go of the fork; it fell. I let go of the two balls; they fell at the same rate. If I, or anyone else, did either or both again, we'd get the same result. There exists no kind of proof of this repeatability. It just seems to be true. If it were not, you can imagine what the world might be like! The fact that nature's behavior *is* repeatable is indeed remarkable and gives us the hope that we can make sense of it. In fact, one of Albert Einstein's lesser known, but profound, remarks is that the most un-understandable aspect of nature *is* its understandability. That is, humans can concoct relatively simple models whose predictions accord precisely with nature's behaviors, which seem endlessly repeatable. Others, in particular the twentieth-century physicist Eugene Wigner, have added to our appreciation of this strangeness aspect by pointing out the unreasonableness of nature's behavior being describable so accurately by mathematical equations. Given nature's very subtle and complicated aspects, it is remarkable to many scientists that relatively simple mathematics does so well in predicting nature's behavior.

This aspect of repeatability implicitly refers to all parts of nature's behavior, which we can repeat and repeat and repeat, like endlessly picking up and letting go of our fork. Here we have confined our discussion to large-scale, macroscopic objects; there are also small-scale, submicroscopic objects, which have special "quantum" behaviors, into which we shall not here delve. If we did, we'd uncover how really strange nature is now known to be!

There is one major exception, a part of science that defies our ability to repeat it even once, let alone ad nauseum. At least for now. I refer here to large-scale events. Some randomly chosen examples are the formation of the earth, the origin of life, and the climate one billion years ago on Venus. For these kinds of events, we can develop models that are consistent with our, usually very limited, knowledge of them. But we cannot now make detailed comparisons or check on such comparisons via repetitions of the events. This problem applies especially to events that took place in the distant past.

There is also an aesthetic aspect of science. For example, successful models, in their mathematical formulations, seem to scientists to have a beauty and elegance about them. Pleasing but puzzling!

What Can You Expect in This Book?

The order in which we shall discuss topics in a major area will be mostly historical. Specifically, for each of the book's three major topics—unveiling the universe, the earth and its fossils, and the story of life—and for many of the subtopics under each, we will start from (near) the beginning to provide a broad perspective on how humans explored various aspects of their universe. In taking this historical approach, we will thus travel back and forth freely in time; I will always try to point out temporal road marks so as to prevent confusion on the ordering of events.

I will share stories about some scientists, not just mention their names as if they were nothing more than paragons on pedestals. By this means, I hope that you will get a feeling for scientists, not as remote gods, but as people, with their blind spots, and other limitations and foibles, which they all had and we have, albeit different in detail.

Very often there is competition between independent scientists and between groups in trying to solve basic scientific puzzles. This is a surprising, if somewhat contradictory, aspect of science research: scientists both compete *and* collaborate, often simultaneously. We will also note the false routes sometimes taken by scientists, leading to dead ends. Near the opposite end, we will often highlight the frequent and fruitful role of serendipity.

Of necessity, the coverage of science in a short book must be both woefully incomplete and shamefully shallow. But my goal is to provide an overview with sufficient detail in selected areas for you to develop an appreciation for both the importance of science to our civilization and the amazing progress made over such a relatively short span of time in our development of science. If you carefully analyze your everyday lives, I will bet that you will be astonished at how many of your activities depend on advances made in science, as I noted at the start of this introduction.

With this awe in the forefront of our minds, let me point out another emphasis in this book: the close connection between science and technology. Science usually advances with technology and vice versa. New technologies, especially nowadays, are enabled by advances in science; similarly, new technologies often enable deeper probes into nature's behavior and hence lead to scientific advances. This tight coupling between the two is often the basis of the changes that affect our lives so pervasively.

Despite the survey character of this book, in what follows I stress *why* we believe what we believe, not just *what* we believe. That is, I stress the basic evidence that leads to each description of nature's behavior and the corresponding model; I don't just present the description and the model. The just-so story approach to science teaching is not my cup of tea. Alas, I cannot always present the basic evidence, but I will make a strong effort to do so. When I can't, I will try to remember to note that fact and its rationale.

Mathematics, as I have noted, plays a key role in science, not only in expressing models of nature's behavior, but in analyzing observations (data) as well. In view of the likelihood that many of you had an education that did not stress mathematics, my approach will be to use mathematics sparingly. Also, where simple mathematics can be used just as well as a more complicated form to explain a scientific concept, I will opt for the simple. Further and perhaps most important, the text is structured such that if you skip *all* of the mathematics, you can still understand almost all of the book quite well.

In places, I will need to use a scientific concept that may be unfamiliar to many of you. In such cases, I try to develop the necessary background to allow you to understand the application to the subject we're discussing. Many of these developments can be considered illustrations of the unity of science because many of these tools are each used (creatively!) in at least a few of the different disciplines of science.

I will *stress* evidence (the basis for what we believe about nature's behavior), and I will *downplay* jargon. I will use everyday words wherever feasible.

You will notice that I frequently use the word "apparently"; I mean to imply by such usage that what follows seems to be correct but may, in fact, be somewhat or even totally incorrect.

What Should You Learn from This Book?

What is the most important message that I would like to deliver via this book? Not simply an appreciation of science and its massive contributions to our advances as a species, but rather something important for individuals, as they proceed with their lives and careers: an appreciation of the importance of *posing key questions*, a skill useful in every sphere of life. I will thus emphasize the asking of questions and the thinking behind the sense(s) in which they are important.

Questions are a major means of stimulation of advances in all areas of our existence. Will we ever run out of them? One impressive consequence, at least I think it so, of such advances is that the more we learn, the more new questions are spawned. We appear to be diverging, not converging, in the number of questions that follow from what we have cumulatively learned at any given time—a seeming paradox. Thus, with the asking of new questions about nature, as we learn more of its secrets, we cannot be sure about finiteness. This onion may have an infinite number of layers!

One more part of the message: I would like you as a matter of course to probe for the evidence behind the inference, to seek out the basis for any conclusion with which you are confronted, always asking why you should believe what you are asked to believe. Also, be skeptical. Try to make sure that the evidence is solid and that the inference follows from the evidence. This is not always easy, or even possible, to do. But you should try.

In addition, given the ubiquitous presence of science in our modern world, I attempt to give you a broad, if not deep, familiarity with science and how it has developed over time, especially its interconnections. Will this background enable you to choose wisely between various policy options with scientific components that often confront the nation and the world? Most likely not, for a simple reason: it would be extraordinarily time-consuming, as well as very difficult, for you to become an expert on even one such issue. It is only a hope that, given some background in science, you will be able to make wiser choices by obtaining answers to key questions and critically examining these answers. A hope more likely to be fulfilled is that you will develop a better sense of the scientific enterprise, its strengths, and the obstacles it faces.

To end this introduction on a light-hearted note, I point out what I consider to be the main difference between the practitioners of science and of magic. Although science may often appear to be magic, there is in fact a profound difference between the two disciplines, and between scientists and magicians. Scientists are almost always delighted to talk to you, and in detail, about their work and discoveries. In fact, they may talk your head off, as I do in a sense in this book. With magicians, it is exactly the opposite. They do demonstrations, too, as I have done and will again do here: consider my stuffing into a cloth bag three pieces of colored cloth securely tied to each other. Then, following suitable incantations, I remove the colored pieces of cloth from the bag. Voilà! The pieces are now untied. I explain nothing. Magicians devote their voices solely to their patter; based on their oath, they never, ever disclose to a nonmember of their formal fraternity how they perform their magic tricks. They take their secrets to the grave and we will return to science.

Unveiling the Universe

ONE

Motions Seen in the Sky

Close your eyes and imagine yourself in a time many thousands of years ago. What would you notice? What would you think about the sky? Why would you care? In addition to your earthly environment, which would be fairly constant in its seasonal cycles throughout your lifetime, there would be (moving) lights in the sky. Most prominent would be the sun, next the moon, then the stars, and finally the planets—"wanderers" in Greek. Continue to place yourself back in those days. There were no knowledgeable people or books to tell you what these objects were. They just *were*. As I noted in the Introduction, that's the secret: nature is what it is. We have no idea why, in a fundamental sense. The goal of science is to probe nature's behavior and to create with our minds models of this behavior. For what purposes? To what ends? In addition to satisfying a seemingly innate curiosity, there existed more practical purposes, too. For one, food: When is the best time to plant crops? When is the best time to hunt game? How accurately does one need to know these answers? And in these questions and answers, what were the relative motivations for practicality, personal interests (for example, religious and astrological beliefs), and plain curiosity? No one knows; we can only speculate. I will forego this pleasure.

People doubtless noticed recurring patterns in the sky. Well below the Arctic Circle in the northern hemisphere (and at southern latitudes well above that of the Antarctic Circle), the most obvious pattern was the day: the regular alternations of light and dark periods when, roughly, the sun was, respectively, visible and not visible in the sky. For the night, keep in mind that there were no distractions, aside from you-know-what. The splendor of the night sky was far more obvious then than it is today. Then there were no streetlights; there wasn't even any smog. Except for possible clouds, fog, rain, or snow, the sky was clear and bright. People looked up and the more curious among them noted changes from night to night, from month to month, and from year to year—they were the scientists of their day, although that name had not yet been invented.

No tools were available to aid such observations of the sky, certainly no telescopes, only the unaided human eye. Measurement accuracy was therefore grossly inferior to present standards but more than adequate to draw dramatic conclusions. Exactly which conclusions were drawn, and in what order, we can only guess. But we can make plausible guesses, which we now do. Observing sunrise and sunset, times when both the sun and some stars could be seen virtually simultaneously, some people certainly noticed, as you for good reason probably have not, that the sun passed each year near the same stars, which formed a band on the sky, with the sun passing them on its annual trip. People may well have more easily noticed this progression near midnight throughout the year (I'll shortly discuss this observation in more detail). The time taken for the sun to complete one cycle through the stars is one definition of the year. In those early days, however, a year was likely marked more by repetition of seasons in temperate northern (or southern) latitudes. It was also quite obvious that the moon's phases had a pattern to them, what we now call a monthly pattern, from new moon (when the moon and sun are both on the same side of the earth) to full moon (when the moon and sun are on opposite sides of the earth) and back again to new moon. Both of these patterns depend on the sun, one directly and one indirectly. One type of new moon event apparently confused and frightened ancient peoples: eclipses, especially total eclipses of the sun that plunged parts of the earth into darkness at midday. Only a total eclipse likely created such fear—even 98 percent of totality wouldn't have been a big deal because our pupils compensate for decreasing light levels by enlarging (dilating). A total eclipse, though, is definitely a big deal. Pursuit of this topic is likely best left to more specialized venues, and we thus so leave it (though we shall revisit eclipses in chapter 3).

The next jump in pattern recognition, we speculate (there are no written records as far as I know), was a big one: periodicities in planetary motion. Planets moved (slowly) with respect to each other and with respect to the stars. The stars, by contrast, did not seem to move at all with respect to each other. The stars were clearly seen to move as a group, every night rotating around the sky; the planets moved relatively slowly on the sky with respect to the fixed pattern of stars and with respect to each other. Extensive observations, perhaps well developed before history began to be recorded, showed that the planets, somewhat like the sun and the moon, had repeatable, albeit complicated, motions among the stars. It was altogether very puzzling.

Some of the first, if not the first, systematic recordings of the positions of objects on the sky were made by the ancient Babylonians, who lived in the land between the Tigris and Euphrates Rivers in what is now Iraq. These recordings started in about 800 BCE—that is, nearly three thousand years ago. The recording medium was clay tablets, and the writing, deciphered in the mid-1850s and credited to Edward Hincks and Henry Rawlinson, working independently, is called cuneiform, which means wedge-shaped. What reference frame, what calendar, and what clock(s) were used by the Babylonians are all pertinent questions, but I, alas, do not know the answers.

The recordings of the positions of astronomical bodies with respect to the stars were made in a rather compact and efficient manner. This mode had another benefit to civilization most likely not realized at that time: about 150 BCE, at least one person from Greece, possibly Hipparchus, visited Babylon and took back with him copies of observations from the previous five hundred to a thousand years, to mark a factor of two uncertainty. These observations were crucial in allowing Greek savants of that period to draw many impressive conclusions about the motions of heavenly bodies. I mention just one: accurate orbital periods of planets, about which more later.

Calendars

Aside from satisfying the curiosity of a few, of what earthly use were such observations? Besides the practical issues mentioned earlier, there were annual celebratory events that people wished to track accurately. These desires in total led to the creation of calendars, or so I think.

In creating calendars—that is, codifications of patterns in the sky—which were the patterns of importance that had to be included? Front and center was the day, the alternations of lightness and darkness. Are these constant? No, and the ancients were quite well aware of the changes. In middle latitudes, the longer periods of lightness occurred during the part of the year we now call summer (hotter) and the shorter ones during the part we call winter (colder). These annual changes may have seemed strange, but that was the way nature behaved. Ancient peoples doubtless noted as well the correlations: when the days were longer, the sun rose relatively higher in the sky at noon and vice versa when the days were shorter. The month, as already noted, went through periodic phases. The year was related not only to the cycling of the seasons but also to the sun's periodic movements with respect to the fixed stars, each yielding a slightly different length year, as we implied above.

All well and good. But how did people produce a calendar? Not easily. And what, you might wonder, was the problem? To use a mathematical term, the root of the problem was *incommensurability* (no common measure). The way that nature was behaving, the year did not contain an integer (whole) number of months (there were about 12.5), nor did the year contain an integer number of days (there were about 365.24). In both cases, too, the relation was not a simple fraction, such as 365¼ for the year when measured in days. (Be aware, too: the lengths of both the day and the month were shorter in the past; due to tidal interactions between the earth and the moon and sun, both the day and month become longer as time passes—up to a point! We will treat tides briefly in chapter 22. For now, please take this effect of tides on faith, sadly a just-so story.)

These facts of contemporary nature spelled doom for the concept of a simple calendar. Let's see why. Suppose that the sun's motions were the most important that we wanted to capture in choosing an annual calendar, composed of days. The best we could do would be to choose a calendar 365 days long, to (most nearly) match a year. What would happen were we to adopt such a calendar? The calendar would, in effect, drift. A given calendar date, for example, January 1, would march backward through the seasons at the rate of nearly a quarter of a day per year. Thus, after about four hundred years, January 1 on the calendar would occur about one hundred days earlier in the "real" season—on days that were near the beginning of fall. This concept is not so easy to understand. Think about it and, if helpful, invent an example with simple numbers and work it out.

Similarly, human events that were tied to particular times of the real year—that is, nature's year (the annual motion of the sun)—would, according to this type of calendar, drift with respect to the seasons as well. What is the solution? There are many possible. One solution, first conceived of, I believe, at least a few thousand years ago in Egypt, and which we still use in modified form, is the insertion of a leap year. In a leap year, an extra day is inserted into the calendar to account for the integer number of days, 365, used to represent the year, not being accurate, but being too short by about—not exactly!—a quarter of a day. Thus, with a leap year defined to be every four years, the drift would be much smaller and of the opposite sign, as the year is slightly less than 365.25 days. To accommodate for this remaining small drift, some leap years are omitted in our present, so-called Gregorian, calendar, named after Pope Gregory XIII. The ordinary leap years skipped—that is, omitted—are the century years whose digits are not divisible evenly by four hundred; for example, 2100 will not be a leap year, but 2000 was. This tweaked calendar also drifts, but at such a slow rate that it is of little practical concern, likely for the order of ten thousand years into the future. In other words, future needs for tweaks to our present calendar do not pose our most pressing world problem or even our grandchildren's!

We have so far neglected the moon, a body of prime importance especially in some religions. Here the incommensurability is different from that encountered between the day and the year. There are neither an integer number of days in the month nor an integer number of months in the year. So, for those civilizations or parts thereof for which the moon was the preeminent body, with its variation in phase being dramatically visible, different solutions were used over time. For example, in some early calendars, there were twelve months, each of thirty days' length, in the year. The drift here of the seasons through the calendar was about twenty times as rapid as in the calendar first discussed above. A solution adopted by some was to stuff five or so extra days at the end of the year, not tied to a month. This solution also had problems, which had to be dealt with and were, differently by different societies.

The history of the development of calendars throughout the world is fascinating. About two thousand years ago, March was the first month of the year, giving rise to September, October, November, and December being, as indicated by their names, the seventh, eighth, ninth, and tenth months. The Arabic, or Muslim, calendar is based on the moon but does not introduce leap days or months. Rather, it lets the months slide through the seasons; for example, the ninth month, Ramadan, appears, over time, during each season. The Chinese calendar is the one in use the longest and has various periodicities, such as a sixty-year cycle, built in. The Hebrew calendar, also lunar based, has a very intricate set of rules, one upshot of which is that the special holidays, such as Passover, move across neighboring parts of nature's year but don't stray *too* far from the intended part. The Mayan calendar has received relatively recent notoriety because it was only carried through December 21, 2012 (on our calendar) and led some to believe that the world would then come to an end (just) because the Mayans hadn't extended their calendar beyond this date. The oldest known calendar is sort of an analog of Stonehenge (massive stones in England widely believed to be arranged so as to indicate such important astronomical events as the first day of summer) but located in Scotland and dated to be about ten thousand years old.

Because of the globalization of trade and travel, most of the world has adopted a single calendar for use by everyone. Our present, Gregorian, calendar, introduced in Europe in 1582, prevailed as the world standard, probably in large part due to the colonization and/or subjugation by European states of peoples in many other parts of the world, along with the increase in global commerce. The last large country to adopt this calendar was Turkey, in 1927. There were complications, as usual: China adopted the Gregorian calendar officially in 1912, but warlords throughout China had different ideas, and it apparently wasn't until 1929 that its use was widespread throughout the land. Switching calendars was always somewhat traumatic at first because it entailed altering dates and thus abruptly changing the days in the year for some customs and practices.

The Week

You may have noticed that we so far completely omitted mention of one major unit of time considered in our civil calendar: the week. Where did it come from? The day, month, and year are rooted in astronomical phenomena. But the week? It has no known relation to such phenomena. It is discussed in the Old Testament, which according to present scholarship has roots back to about 1,400 BCE. There are also substantial indications that the week was used by the Sumerians as long as four thousand years ago. The week most likely was invented earlier for reasons about which we can only speculate. Try as I did, I could find no reliable evidence on the origin of the week nor on its now being seven days in length. I did, however, learn that a week of seven days, with names, as ordered now, was used in ancient Babylon. I trust that we'll learn more in the future about our past; some may know more now.

The length of the week, apparently, did go through different values at different times in different places. Why it is now roughly one-quarter of a month, as opposed to, say, one-fifth of a month, is less than clear to me, although it may be that a seven-day week roughly matches the duration of major phases of the moon: new moon to half moon, half moon to full moon, and so on. But the names of the days of the week, listed here in English (from Old English using Norse gods) and in French (from Latin), do provide an intriguing hint that may or may not be relevant (also, it's not clear to me which is cause and which effect!):

Sunday/Dimanche
Monday/Lundi
Tuesday/Mardi
Wednesday/Mercredi
Thursday/Jeudi
Friday/Vendredi
Saturday/Samedi

Each name is related to an astronomical object, in order: the sun, the moon, Mars, Mercury, Jupiter, Venus, and Saturn. These names encompass all of the known heavenly bodies in ancient times, except for the stars and our own platform, the earth. (Can you guess why the earth was omitted from this list?) Why the days are ordered this way is less clear, except possibly for the beginning of the week. But the number of these objects may have been the reason for the seven-day week.

It is also true that even though there is no integer number of weeks in the month or in the year, nobody seems to care. We could argue that this fact keeps us on our toes, as from one year to the next each date occurs on a successive day (except for leap-year oddities, when in general we jump two days of the week as a consequence; can you think why?).

Over the years, many people have noted patterns, for example between the days of the week and the twenty-four hours in the day (another interesting story, omitted here, with unclear details on origin). These seem to me to be little more than playing with numbers, without any obvious or clear significance. A similar statement applies to the subdivisions of a day: hours, minutes, and seconds, whose origin seems to have been in Babylon but whose impetus is less clear, except to note that their numerical system was based on the number 60 instead of, as for us, the number 10. (If you're curious to learn more, you may, for example, consult the Web.)

The Solar versus Sidereal Day

Having noted that the solar day is at the heart of the calendar, I would like to point out the difference between the solar (= sun) and the sidereal (= star) day. Let us delve into this relation quantitatively—namely, the connection between the number of days in the year as measured by sun days and by star days. These two types of days are different, such that the number in a year of star days is one more than the number in a year of sun days. Why is that? The answer lies in how the earth moves with respect to the sun compared to its motion with respect to the stars. Observations—all of course made from the earth—show that the sun moves around the sky—that is, with respect to the fixed stars—exactly once per solar year, by definition. What does that mean? Let us explain: The day as measured with respect to the stars is the time it takes between a star crossing your meridian (the line or plane of longitude that passes through your location) and when it next crosses your meridian. This so-called sidereal day is slightly shorter than the day as measured with respect to the sun (the time elapsed between one noon and the next noon, where "noon" is defined as the instant during a day at which the sun crosses your meridian), the so-called solar day. We have these days because the earth spins on an axis, but the earth also travels around the sun at a rate of approximately 1 degree per day. As a result, in order for the sun to cross your meridian the next time, the earth must spin that extra one degree, which takes about four minutes. But because the star is so much farther away from us than the sun, that extra one degree of rotation is not required for the star to cross your meridian again. If you add all those four-minute blocks up in a year, they mean that there is one more sidereal day per year than there are solar days. An illustration drawn from a modern, heliocentric (sun-centered) perspective (fig. 1.1) describes this situation. From this drawing, we can see that from the perspective of an observer on the earth, the sun seems to move in front of the stars, so a solar day will be a little longer than a sidereal day: the earth's rotation has to catch up with the apparent (to us) motion of the sun with respect to the fixed stars. Thus, in the left panel, the arrow points directly overhead toward some particular star—from some location on the earth's surface, say a point on the equator. In the right panel, drawn for the situation one sidereal day later, the same arrow points to the same star, but the earth has moved in its orbit by one day, roughly one degree as measured from the sun. The earth in its daily rotation about its (polar) axis has to move about one degree more for the overhead pointing arrow again to point toward the sun, marking a solar day. We see thereby that the solar day is longer than the sidereal day. *There are thus fewer solar days than sidereal days in one year, in fact exactly one fewer.*

Note that both panels of the illustration are drawn in a frame of reference at rest with respect to the fixed stars. Note, too, that we have, in effect, ignored the third dimension in the figure and in our argument, for simplicity; the same ideas would hold were we to carry out a full three-dimensional argument. Put the two-dimensional argument another way: the straight arrow in the left part of the figure points to the sun and certain stars; one sidereal day later, the earth—right part of the figure—points to the same stars, but

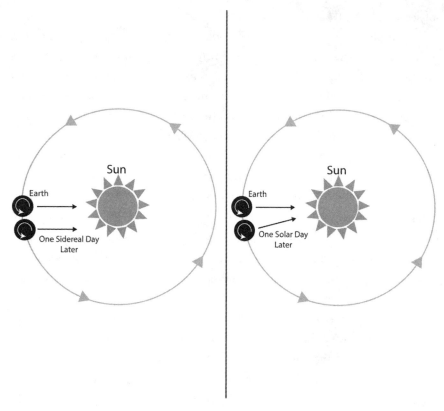

Figure 1.1. The difference between a sidereal and a solar day from our modern perspective. Courtesy of David Shapiro.

due to the earth's movement about the sun (remember: the drawing is from our modern perspective), the arrow no longer points toward the sun. It takes a little longer for the rotation of the earth to catch up to the sun. Thus, the solar day, as stated, is a corresponding time longer than the sidereal day. In one year, the sun has made one complete revolution with respect to the stars, thus leading to one less solar than sidereal day in a year, as I hope you can see by extrapolating from the illustration to a complete year.

I note again that we have interpreted this difference between sidereal and solar days with a drawing based on our current sun-centered model, not from the point of view of the ancients, which was earth-centered. We could, however, have used their viewpoint of the earth standing still and reached the same conclusion.

Let us amplify briefly this intriguing notion that the sun drifts among the stars. How could we or, more to the point, the ancients know this? Everyone understands that stars can be seen only when the sun is not visible; stars cannot be seen when the sun is visible (and vice versa!). How then can we draw such conclusions about the sun's movement with respect to the stars? As mentioned above, as the sun sets or rises, we can glimpse some bright stars before either the setting or the rising is completed, just at either's start. Better yet, we can observe the motions among the stars of the successive full moons. Some ancients must have figured out that during the full moon, the sun must be in the exact opposite part of the sky, as seen from the earth. They knew this because they had doubtless already figured out that the phases of the moon were caused by the moon shining only by reflecting light from the sun. Thus, observation of the full moon's cyclical progression relative to the star background indicated that there was a similar cyclical (annual) progression of the sun's positions among the background stars. These stars were labeled by the various constellations—sets of stars named in antiquity, based on imaginative associations of their shapes, as seen in the sky. A clear implication of this drifting is that the number of days in the year depends importantly on the definition of a day, as we discussed. Do we reckon by the sun or by the stars? We reckon by the sun.

TWO

Modeling the Motion of Celestial Bodies

With our discussion of calendars now as complete as we will make it, let us turn to the attention ancients paid to those other-than-stellar points of light in the sky. The motions of the sun and moon were blatantly obvious, due to their sizes, brightness, and relatively simple periodicities of motion around the sky. Ancients also devoted attention to those other moving points of light, the five then-known planets. Their motions were rather complicated and harder to model accurately, as we'll see. One aspect, though, that the ancients were aware of millennia ago was that all of these objects moved in nearly the same plane—the whole system was fairly flat.

The earth was not then recognized as a planet. Why not? Try to formulate a concise answer that could convince a skeptic.

In general, humans had the urge to predict the positions of planets in the sky versus time. This urge was likely prompted in part by curiosity and perhaps in larger part by the astrological thinking of the day—namely, that planets' movements and positions played a critical role in people's lives. But people of that era had no idea on what to base predictions of planetary positions. They developed what we now demean as ad hoc models to be as consistent as they could make them with the data. (A model not based on a principle but rather designed to deal only with the relevant data is usually called ad hoc.) These people did not succeed too well. There were, for a prime example, severe difficulties in predicting so-called retrograde (backward) motion of planets on the sky (fig. 2.1). In the schematic picture

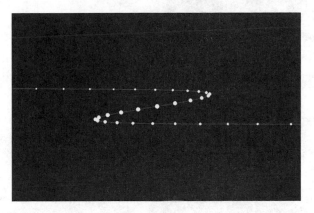

Figure 2.1. The retrograde motion of Mars. Courtesy of David Shapiro.

shown here, Mars is seen at intervals of about a week, starting from the point of light (white dot) in the lower right and moving to the upper left (west to east), starting out moving relatively rapidly (large spaces between dots), slowing to a halt (bunching up of dots, which are close to equally spaced in time), then reversing direction on the sky (the retrograde part of the orbit), again speeding up and slowing to a halt, and then again reversing direction and again speeding up (dots farther apart) to the upper left of the figure. Why the retrograde part of the orbit occurred at all, why it occurred where in the orbit it did, and why it lasted as long as it did were all mystifying features of planets' motions on the sky, especially that of Mars. It was another very puzzling aspect of nature's behavior.

Building on the knowledge of earlier civilizations, the Greeks created some of the earliest models of happenings (paths of objects) on the sky of which we have any fairly direct knowledge. Aristotle (384–322 BCE), and probably his teacher Plato and other predecessors, envisioned a universe with the earth at its center, with stars affixed onto a transparent crystalline sphere that rotated at a uniform speed about the earth once per day. Such behavior was considered by Aristotle and contemporaries to represent perfection; their model also included other, similar crystalline spheres, rotating uniformly, one each for the sun, the moon, and each planet. On the earth, objects fell to the ground because "of course" objects tried to fall to the center, which was obviously the center of the earth. Also, it was clearly obvious that the earth wasn't moving: Who standing—or sitting—on the earth could sense that they were in constant motion? No one. And certainly no one then, as opposed to today, had any evidence for such motions being up to 0.5 kilometers per second about the earth's polar axis and of 30 kilometers per second around the sun!

A reminder: our purpose in reviewing these historical models is to provide perspective on how truly difficult it was to produce models consistent with nature's behavior as exhibited in the observed motions or paths of celestial bodies. Even though, from our modern perspective, we are tempted to treat their struggles and models with bemusement, they were in fact major triumphs for their day, requiring the finest minds of the time to conceive and develop.

Two Greeks, Apollonius of Perga (ca. 240–190 BCE) and Hipparchus (ca. 190–120 BCE), are generally credited with taking the first steps toward a comprehensive model of what we now term the solar system, a model *based solely on observations*, then a new concept. Hipparchus, as opposed to Apollonius, had the further new idea that the model should match the observations *in detail*, rather than just in a general way. This idea represented an important advance in scientific thinking. Of course, it was then quite clear that Aristotle's model didn't come close to matching the observations, especially for the planets.

One thing Hipparchus, among others, noticed was that the sun did not move on the sky with respect to the stars at a uniform rate throughout the year. In the (northern hemisphere) winter, the rate at which the sun moved across the sky in angle (its angular rate) was noticeably faster than during the summer. Thus, the winter season was observed to then be about eighty-eight days and the summer season about ninety-three days in length. So Aristotle's model was clearly at variance with observations. What to do? Hipparchus (or perhaps Apollonius earlier, or even someone else—we really do not know) had the excellent idea of preserving the uniform angular rate part of the earlier Aristotelian model by moving the center of the sun's circle away from the center of the earth by just the amount and in the direction needed to agree reasonably well with the observed length of the seasons as viewed from the earth. Keep in mind that when one is observing an object moving uniformly around a circle, the closer one is to the object the faster, angularly, it appears to be moving; similarly, the farther one is away from the object, the slower it appears to be moving in angular rate. Remember, too, that of the pair, we think that only Hipparchus believed in the need for the model to agree with the observations in detail.

Even though it clearly disagreed with the observations, Hipparchus tried to preserve the basic Aristotelian model by maintaining the key concept of uniform motion around a circle, while moving the center of the sun's circle away from the center of the earth. This is an important aspect of science as it is usually practiced: move gradually away from your current model in making improvements; dramatic breaks are usually too radical. However, dramatic breaks, though far less common—we'll see several in this book— are usually where the biggest payoffs are obtained.

Ptolemy's Model

At this point, our history turns to Claudius Ptolemy (ca. 90–160 CE), who lived in Alexandria, Egypt. His first name shows the Roman influence there at that time. The Roman Empire then encompassed Egypt as well as much of the Middle East and Europe.

Ptolemy devoted a large part of his life to producing a book he called *Syntaxis*. This book was renamed by Arab scholars, who made the main advances in science somewhat after that period; the book is now universally known under their name for it, the *Almagest* (The Greatest). The *Almagest* contained predictions in the form of astronomical tables that allowed people to determine where in the sky they could see each known planet at any chosen time. Times were based on the Julian calendar, introduced under Julius Caesar in 45 BCE. Ptolemy first needed a model on which to base these predictions. He chose Hipparchus's model, in which an epicycle (a jargon word, meaning, in general, a smaller circle whose center rolls on the circumference of a larger circle at a constant speed) was introduced to deal with retrograde motion in the sky (fig. 2.2).

In this model, the planet moved on the circumference of the epicycle around its center at a uniform rate with respect to this center. The center

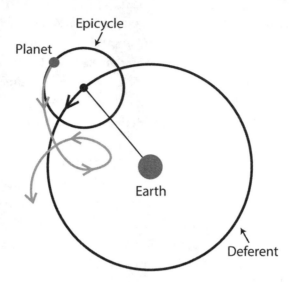

Figure 2.2. The epicycle and retrograde motion (see text). Courtesy of David Shapiro.

of the epicycle, in turn, moved uniformly along the circumference of the larger circle (the deferent—more forgettable jargon!) whose center was off-set from the earth. As noted earlier, the center of this large circle was offset from the center of the earth mainly to account for nonuniformities in the observed motion of the sun about the earth. For example, this offset could account for the different lengths of the seasons, while maintaining the Aristotelian emphasis on the perfectness of nature, which implied perfect circles and uniform motion—motion at a constant rate. As can be seen in the illustration here, in which the planet's movement is shown, the planet (the small black dot) will appear to an observer on the earth to be moving backwards along the bottom half of the light gray loop, the part of the loop closest to the earth. So as not to introduce too many complications at once, we omit in this figure, but include in the next, the offset of the earth from the center of the deferent.

This epicycle approach still didn't do very well in predicting the retro-grade parts of the orbit. In essence, Hipparchus tried to have his model match both the angular extents of the retrograde parts of the orbit and the positions in the sky of these parts of the orbit. Both of these aspects needed to be matched for successive retrograde parts of the orbit. He had found no configuration that worked. Ptolemy thus sought a new idea to better solve this problem of simultaneously matching the widths *and* sky positions of the retrograde motions, especially for Mars. He came up with a clever and novel idea; he introduced the equant (the last bit of jargon for this discussion!).

How do we know that it was Ptolemy's idea? We don't. But since in his book Ptolemy doesn't refer to any source for the idea of an equant but does give the sources for other ideas, everyone believes that the equant was his idea. What exactly is this idea? The idea is to introduce a point about which the center of the epicycle would appear to move at a uniform angular rate, to preserve the Aristotelian idea of uniform motion, while allowing closer agreement with the observations of retrograde motion. The equant was placed on the line connecting the center of the large circle (deferent) with the center of the earth, but on the opposite side from the center of the earth and the same distance from it as is the center of the earth (fig. 2.3). (This

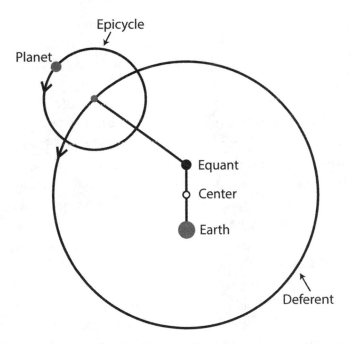

Figure 2.3. The equant, introduced by Ptolemy to better represent retrograde motions. Courtesy of David Shapiro.

arrangement has similarities to an ellipse—see below—in which the foci are located at the earth and the equant, foreshadowing if you will.)

Why did the introduction of the equant improve the agreement of the predictions of Ptolemy's model with the observations? This is not easy to explain without a lot of words. In essence, he introduced a new aspect to the model that allowed him to have simultaneously a feature of the model that gave pretty good agreement with the widths of successive retrograde parts of Mars's orbit and a feature that allowed him to obtain pretty good agreement with the positions on the sky of the successive retrograde parts of Mars's orbit.

Suppose Ptolemy had placed the equant differently, could he have obtained better agreement between his predictions and the observations? Perhaps, though I suspect not much, if any, better. But I haven't tried. There is no indication by Ptolemy whether he tried to improve upon this model and, a fortiori, what, exactly, he had tried as alternate models and what he had found.

With the introduction of the equant, Ptolemy's model was qualitatively complete. But Ptolemy's goal was to create a model and from it tables that could be easily used to determine the position in the sky of the sun, moon, and planets for any chosen time. By converting his qualitative model to a quantitative one in a manner we will skip, he developed very compact tables that enabled the user to determine near which star(s) the planet could be seen at any chosen epoch. Since Ptolemy used observations of planet positions obtained by the Babylonians, extending back to perhaps 800 BCE, the accuracies of the predictions did not degrade rapidly with time and the tables in the *Almagest* remained in use routinely for about fifteen hundred years. The instruments in use to measure the planetary positions in Ptolemy's day were all naked-eye instruments, and the accuracy of these did not improve very much during this time.

Medieval Arabic astronomers made some important improvements to Ptolemy's model, particularly in correcting an error in the rate of movement of the earth's rotation axis with respect to the stars. The earth's rotation axis moves circularly about the line perpendicular to the earth's orbital plane, with a period of about twenty-six thousand years—the earth's so-called precession. This circular motion is modeled as due to the gravitational effect on the flattened earth of the moon and the sun. (See chapter 4 for a discussion of the laws of gravity, although not a discussion of this calculation, which is well beyond the scope of this book. The angle between this rotation axis and this perpendicular line is about 23.5 degrees. The differences that we experience in the seasons are due to the relatively large size of this angle.) Ptolemy had used Hipparchus's value for this rate of movement, which was known by the thirteenth century to be too small by about 30 percent. This rate of change of direction is in total about 50 arcseconds a year. (An arcsecond is a very small angle; it is one-sixtieth of an arcminute, which in turn is one-sixtieth of a degree.) Thus, after this 30 percent correction, the errors in prediction were then reduced to being comparable to the measurement errors of that era.

Although Ptolemy tried with his theory to adhere to Aristotle's (and Plato's earlier) ideas of perfection of the heavens, nature did not cooperate. Even with all of the violations of Aristotle's picture of perfection that Ptolemy introduced into his model, his theory still did not work well in some of its major details: the retrograde rotation of Mars, in particular, provided obvious discrepancies between prediction and observation.

Copernicus's Model

The next major advance in modeling the motions of the planets did not come about until the sixteenth century. In Poland in that era, there was a young man, preparing for the clergy, who was very interested in astronomy. He developed a different model to explain planetary motions. Nicolaus Copernicus (1473–1543) in his model assumed that the earth rotates daily on its polar axis and that the planets, including the earth, all revolve about the sun. Whereas Ptolemy's theory was more or less geocentric, Copernicus's theory was heliocentric to about the same extent. Copernicus's model also resembled Ptolemy's: it had large circles, *not centered on the sun* (in this sense, Copernicus's model was not heliocentric; it was only *nearly* heliocentric), and epicycles, although relatively much smaller than in Ptolemy's model, as well as the same general approach to determine the values of the quantities for the descriptors of each planet's motions. Because, as we now know, planets move in noncircular orbits about the sun, Copernicus incorporated offset circles and much smaller epicycles to obtain reasonable agreement between the predictions of planetary positions from his model and the actual measurements of planetary positions. Copernicus, too, produced tables for determining sky positions of the sun, moon, and planets as a function of time, according to the Julian calendar. The positions of stars with respect to each other formed the reference system used for describing planetary positions. Copernicus apparently abhorred the equant, and it is absent from his model. Apparently, he thoroughly disliked the violation of uniform circular motion about the center of the deferent that the equant entailed and, in any event, managed his own theory quite well without an equant.

Despite the leap forward provided by his (nearly) heliocentric model, calculations of planetary positions using Copernicus's theory were not particularly easier to make than were those with Ptolemy's theory. Moreover, the accuracy of their results was similar. So why would anyone bother with Copernicus's theory? What did it have to offer other than the headache of contravening thousands of years of belief by moving the center of the universe from near the earth to near the sun and thereby possibly running afoul of the Catholic Church in which Copernicus was serving as a canon?

There were three main conceptual advantages of the nearly heliocentric model over the nearly geocentric one. The first dealt with retrograde planetary motion. In Ptolemy's model it was a curious coincidence that whenever retrograde motion of Mars, Jupiter, or Saturn was taking place, the sun would be behind the earth—that is, in the opposite direction in the sky from the planet. In Copernicus's model this alignment is natural. Let us use Mars as an example (fig. 2.4). The earth, having the shorter orbital period (the time for a planet to complete one full revolution about the sun), moves around the sky faster than Mars. When the two are almost aligned on the same side of the sun, the earth goes by Mars in the sky, and hence Mars appears to undergo retrograde motion: the direction of the earth-Mars line first rotates counterclockwise (1, 2, and 3 in the illustration), stops rotating (3), next rotates clockwise (3, 4, and 5), again stops rotating (5), and at last resumes its counterclockwise rotation (5, 6, and 7).

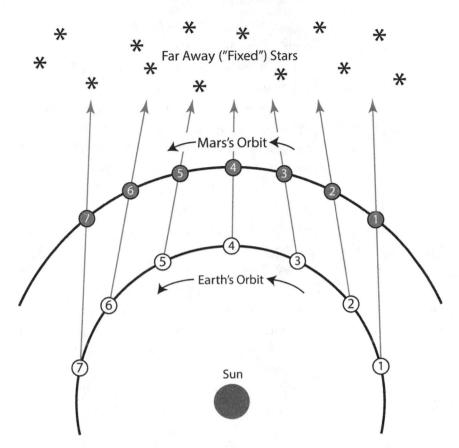

Figure 2.4. The retrograde motion of Mars as viewed from the earth but pictured from above (north of) their (nearly) common orbital plane, at various times (1–7). Note: this figure is *not* to scale; the background stars are, of course, far, far too close, and the distance of Mars's orbit from the sun is a bit more than 50 percent larger than the earth's (see text). Courtesy of David Shapiro.

The second advantage of the nearly heliocentric model deals with the relative distances of planets from the sun. In Ptolemy's system there was no particular relation between the periodicities of the motions of the planets on the sky and their distances from the earth. In Copernicus's system there exists a clear correspondence: the longer the orbital period, the greater the distance of the planet from the sun.

The third advantage explained why Mercury and Venus always appear close to the sun in the sky. In Ptolemy's system, this fact had no explanation.

In Copernicus's system, it was a consequence of their orbits being closer to the sun than is the earth's. Copernicus's theory thereby connected two otherwise seemingly independent facts: these planets' rapid motion in the sky and their staying near the sun.

There was also a fourth characteristic of the planets' behavior explained by the Copernican theory: the changing relative brightnesses of the planets as seen from the earth. In the Ptolemaic theory, these differences in brightness were left unexplained; in the Copernican theory, they had a natural explanation: in general (there are exceptions—can you think of any?), the closer the planet to the earth, the brighter it appeared. This issue was not of great significance in Copernicus's time because of the lack of technology to accurately compare the brightnesses of a single object at different times in different places in the sky. The inverse square law dependence of the apparent brightness of a light source on its distance from the observer was also likely unknown in that era. (Demonstrate the validity of this law for yourself: take a simple photometer [quantitative light detector] and move it to various distances from, say, a flashlight and calculate whether the brightness recorded by the photometer decreased with the inverse square of the distance [$1/d^2$, where d is the distance] of the photometer from the end of the flashlight from which the light emerged. For example, if the photometer is moved twice as far away from the flashlight, the intensity of the light recorded by the photometer should drop to one-half squared, or one-fourth of its previous value were the inverse square law to apply.) But it was well known that the farther away a light source was, the dimmer it would appear, all other factors being the same. Copernicus's model was consistent with that knowledge; Ptolemy's model was not.

We know of this work of Copernicus through his masterpiece, *De revolutionibus orbium coelestium*, published when the author was nearly on his deathbed. Did he delay publication because of fear of the reaction of the Roman Catholic Church? There is good reason to challenge this common belief. Copernicus had two friends, one a bishop and one a cardinal, who urged him to publish. Further, Giordano Bruno, who adhered to Copernicus's theories and was burned at the stake for heresy in 1600, apparently claimed that his heliocentric cosmology was not relevant to his meeting that end. But there is also reason on the other side, as can be inferred from the defensive tone of the introduction to Copernicus's work, which implied that there was opposition to it. We leave to historians the last words on this issue and hope that they can both agree upon them and get them correct. In any event, the church did not ban Copernicus's magnum opus until 1616 and then did not remove the ban until 1835.

How was this radical Copernican theory received in the Western world? Words such as "heresy" and "blasphemy" commonly peppered reactions to the new theory. Copernicus had proposed a truly profound change in the world system, demoting the earth from a preeminent position in the universe to being a mere participant in a system presided over, or controlled by, the sun. As we discussed above, despite its advantages, the Copernican system was not without its flaws. It's hard for us to appreciate these reactions since, from our vantage point, it's obvious which model is closer to being correct, and we're very used to our—the earth's—downgraded status.

Improved Observations: Tycho Brahe

The next big advance was observational. In the last half of the sixteenth century, there was a very energetic, charismatic person passionately devoted to observations of heavenly bodies, especially planets and most especially Mars: Tycho Brahe (1546–1601). He obtained financial backing from the king of Denmark, who underwrote the cost of an entire observatory at Uraniborg, on the then Danish island of Hven, replete with instruments designed by Tycho—still all naked-eye instruments, but with which he recorded far more accurate measurements than had preceding astronomers. His measurements of angles had uncertainties as low as one minute of arc (one sixtieth of a degree). He was indefatigable as an observer. Over the course of about two decades, he recorded more than four times as many planetary observations as had then been made throughout the whole world together. He especially accumulated observations of Mars, and particularly during and near its periods of retrograde motion.

Tycho also developed a new model of the solar system, a hybrid between the Ptolemaic and Copernican models. Its main characteristic is that the planets revolve around the sun but that the sun and the moon revolve around the earth. The earth remains at the center of the universe in Tycho's model. This model was motivated in part by the fact that no one had observed annual motion of stars on the sky as would be expected if the earth moved around the sun (see the discussion on parallax in chapter 5 to learn why such annual motion had been expected but had not been observed). This model, though, had difficulties and was not considered seriously by many scientists, although it was a clever, motivated approach to salvaging a geocentric universe.

Whereas Tycho devoted enormous effort to creating his observatory, designing his excellent instruments, and making extremely accurate and numerous observations of planetary positions, he did not spend much time analyzing them. He needed someone who could do this analysis—a forerunner of the need for cooperation among different specialists to advance rapidly in science. He found this person in Johannes Kepler (1571–1630), a young German mathematician.

An Improved Model: Johannes Kepler

Kepler received an appointment from Tycho to work as his assistant in 1600, after Tycho had accepted a position with King Rudolf II in Prague. (Tycho's patron in Denmark had died, and Tycho went to where he could continue to obtain support. Not stupid, he.) Kepler was obsessed with the difference between the observed and predicted positions of the planets, especially of Mars during its retrograde periods. He was equally obsessed—not that I could really quantify these two obsessions!—with the idea that there had to be a physical reason behind the motions of the planets.

Why did Kepler not think that the model of Copernicus possessed such a reason? In Copernicus's model the sun was *not* at the center. The large circle was offset, as was the epicycle; although in Copernicus's case, the epicycle was relatively small. Kepler felt that the sun should be the seat of the physical cause of the action. He strove to develop a model that would have that attribute and closely match the observations—a tall order! First, he had to solve a basic problem: access to the data. Tycho died less than a year after Kepler came to work as Tycho's full-time assistant. There was a long, drawn-out battle with Tycho's heirs for access to his data, after his untimely death in 1601 at age fifty-four. Until recently, it was thought that Tycho died of mercury poisoning from a (mostly) silver partial nose replacement, for the part of his nose lost in a duel fought in his youth with a distant cousin over some mathematics question. After recent studies of his exhumed body, it was found that his nose prosthesis was apparently made of brass, which contains no mercury, and that he died rather from a bladder problem said to have been caused by his not wanting to leave an important dinner to relieve himself. This is yet another tidbit of history that had an important effect on scientific progress, even if it may not be (entirely) true!

Kepler eventually gained access to the data from Tycho's heirs and tried to develop a model of planetary motions that was manifestly attributable to the sun's actions. He thus broke from offset-circle tradition. With inspired and exhaustive (as well as exhausting) trial-and-error work, Kepler created a brilliantly simple model. No more offset circles, no more epicycles, no more equants. And, of course, no more disparities between observations and predictions for retrograde motions. His model could be encapsulated in what are now universally called Kepler's three laws of planetary motion. (Though recall in the Introduction our warning regarding laws in the context of science.) These laws were profound insights for their time. Moreover, they satisfied Kepler's goal of seeking an explanation in which the sun itself was central, not an offset circle. Kepler's thoughts on why the model worked were, however, off base. He thought that the cause was magnetic, perhaps influenced by his contemporary William Gilbert's conclusion in 1600 that the earth resembled a big (bar) magnet. There was no known reason then — nor is there today! — to think that *magnetism* was related to planetary motion; it was presumably just a hunch that Kepler had that turned out to be incorrect. However, magnetism was a force that acted at a distance, and this fact may also have contributed to Kepler's choice.

As eventually formulated, Kepler's laws were remarkable in their simplicity and their accuracy in predicting planetary motions; it whetted the appetites of many scientists for a "deeper" theory. (Why do I insert quotation marks here? Because the concept of "deeper" is partly a matter of individual taste and not an issue of mathematical proof based on agreed-upon principles.)

Kepler formulated his first two laws by 1609; nearly another decade passed before he had finished the development of his third law. However, the modern articulation of Kepler's three laws, which we will discuss, was written more than 150 years later by J. J. Lalande (1732–1807), a French astronomer.

Kepler's first law states that planets follow elliptical, rather than circular, orbits in space, with the sun fixed at one focus of the ellipse. The definition of an ellipse, by the way, can be given as all points in a plane the sum of whose distances to two fixed points in that plane is some specific value greater than the distance between those two points. If that is hard to grasp, use this experiment to visualize an ellipse: take a piece of string of any length, hold the ends down at two points on a piece of paper, with the string sufficiently long that it is then slack. Stretch the string taut, move a pencil inside the string keeping it taut, and trace out a (closed) curve on that piece of paper. That curve will be an ellipse. The two points holding the string are called the foci of the ellipse. The longest straight line that can be drawn inside the ellipse is called its major axis; half that value—surprise!—is called the semimajor axis.

Kepler's second law states that planets sweep out equal areas about the sun in equal times (fig. 2.5). The area swept out means *the area encompassed, or passed over, by the line from the sun to the planet as the planet moves in its orbit.* In the figure, therefore, each of the three dark gray areas would be swept out in equal times by the orbiting planet. The third law affirms that the square of the orbital periods, P, of the planets' motions about the sun are proportional to the cubes of the corresponding semimajor axes, a, of their orbits, with the constant of proportionality being the same number for all planets orbiting the sun:

$$P^2 = constant \times a^3, \tag{2.1}$$

where the units of the constant have dimensions of time squared/distance cubed.

What would the effect be of each planet's having a different constant in equation 2.1? The equation would be vacuous! Why? For any pair of values (P,a), a constant could be found for which the equation would be satisfied.

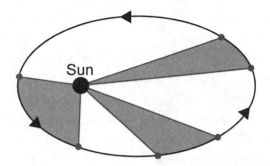

Figure 2.5. A planet sweeps out equal areas in equal times, according to Kepler's second law of planetary motion. The sun is at one focus of the orbital ellipse, here represented by the large black circle. All three gray regions have the same area and represent the same time of planet sweeping. Courtesy of David Shapiro.

Usually, for this constant, units of time in years and units of distance in astronomical units are employed. (One astronomical unit is, roughly, the average distance between the earth and the sun.) Astronomers had little idea of the distances between planets in the units of length used on the earth, for example kilometers or miles. Using these units—years and astronomical units—had a neat consequence, which of course was responsible for their use: the constant in equation 2.1 becomes unity, or 1. Since the earth's orbital period is one year ($P = 1$), and the earth's orbital semimajor axis is about one astronomical unit ($a = 1$), the constant must also be 1 for the equation to be valid. The constant has the value unity ("1") for all planets in our solar system and thus:

$$(P^2/a^3) = 1, \tag{2.2}$$

or, equivalently:

$$P^2 = a^3. \tag{2.3}$$

Kepler's laws are not laws in the inviolable sense, but they do provide rather good approximations to nature's behavior, particularly for the accuracy of observations available at that time, the early 1600s. In particular, Kepler's laws made predictions that were in remarkably good agreement with Tycho's observations of the planets' positions. When applied, for a prime example, to the earth and Mars simultaneously, the glaring errors in the predictions of retrograde motion that had plagued previous models disappeared without a trace.

Kepler's breakthrough was a very significant milestone in the development of modern science. His laws, though independently postulated and not following from any simple single principle, were giant steps forward. From a modern viewpoint, we consider these laws to be ad hoc, but they were, as noted, a brilliant advance in bringing models up to the level of measurement accuracies of the time, the first such advance in about fifteen hundred years of human attempts to quantitatively model planetary motions.

THREE

The Telescope and the Speed of Light

We now turn to discuss contemporaneous, and critically important, astronomical investigations then underway elsewhere. While Kepler was doggedly analyzing data collected by Tycho, a far-reaching revolution in astronomy was taking place about 800 kilometers south. What was it? The initial use of a telescope in Padua by Galileo. The telescope was apparently first used in the Netherlands at the end of the 1500s or the early 1600s; the proper distribution of credit for this incredible invention will likely never be definitively resolved. This is a major detective story, as described by Albert van Helden in a fascinating book, *The Invention of the Telescope*. The main point for us is the invention itself and its use. Originally, in northern Europe, it was used for spying on enemy armies and for other such worldly pursuits. Galileo heard about this invention, built his own telescope, with an optician's help, and turned to its use for otherworldly pursuits — to look up. The first telescope that he built, then state of the art, was exceedingly primitive by present-day standards; its main lens was only 3.7 centimeters in diameter. Its optical properties were very poor: distorted images and fairly low-power, but a reasonably broad field of view. It was also difficult to point the telescope in the direction you wanted to point it, as no mounts were originally available to hold the telescope steady. Nonetheless, the telescope, originally called an *occhiale* by Galileo, enabled him to open up new worlds.

Among the targets toward which Galileo turned his telescope was Jupiter, early in January 1610. Instead of seeing merely a point of light, Galileo became the first human to see that a planet was a clearly distinguishable (nearly) round disk, an object with non-zero size; quite a thrill, I imagine. Then, of course, near Jupiter there were stars in the sky, which Galileo also indicated in his notebooks in their correct relative positions. He made such entries on each of the days he observed. When he compared these various entries, he noted something totally unexpected. Some of the pinpoints of light seemed to move relative to one another from night to night, unlike stars; moreover, they seemed to "stick" to Jupiter, and most remarkable of all, they remained in nearly a straight line, albeit with changing distances between them on the sky and different numbers of them appearing on different nights, with the maximum being four.

How long did it take the master detective Galileo to figure out the proper interpretation of his observations? Approximately a week after he first observed Jupiter. Owen Gingerich, professor emeritus at Harvard University, and a colleague examined all the relevant surviving documentation from Galileo and, using modern computers to reconstruct the appearance of the sky in 1610, were able to determine the likely date of his eureka moment—January 13, when Galileo saw all four objects, not in a straight line. They couldn't be stars, he realized. Instead he figured out that they must be moons of Jupiter. Imagine what this realization meant! Keep in mind that the Copernican system was far from universally accepted in that era. Now, here he finds Jupiter with four satellites circling around it—a miniature solar system, at least in the Copernican worldview, with Jupiter playing the role of the sun. Galileo also realized that the reason the satellites of Jupiter didn't always appear in an exactly straight line was likely that we were not in the same plane in which they were moving; rather, we were looking from above or below that plane. Since the satellites were in ever-changing positions along their orbits, the satellites did not always appear in a straight line, nor were all four always visible, given that sometimes our view of one and/or another would be blocked by Jupiter.

Despite the modest capabilities of his telescope, it was the first ever used for astronomical purposes. Thus, Galileo was able to make other remarkable discoveries as well, including the phases of Venus (fig. 3.1). Note that the (nearly) full Venus occurs in the early evening, soon after sunset,

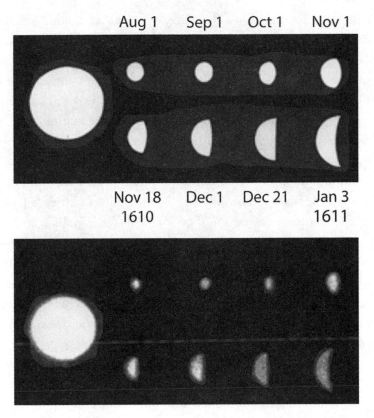

Figure 3.1. Simulations by Owen Gingerich of images of Venus (top) and as likely seen by Galileo (bottom), between mid-1610 and early 1611; at left for size comparison is an image of Jupiter when it is closest to the earth. Courtesy of Owen Gingerich.

when Venus is just out from behind the sun, with successive phases visible as Venus gets closer to earth (and, in the last view, when it is nearly between the earth and the sun, it is seen as a crescent). Notice, too, that when Venus is closer to earth, when it appears as a crescent, it appears in the sky with a considerably larger diameter than when it appears nearly as a full Venus, at which position it appears markedly smaller because of Venus's much greater distance from earth at that point in their relative orbital positions. This was yet another observation consistent with the heliocentric system, but not proof of it (what would constitute *proof* in such matters is alas unclear, as noted earlier).

Galileo also discovered that the Milky Way was not a diffuse cloud but in fact a rather dense collection of individual stars. Galileo may be most popularly known to have been the discoverer of craters on the moon, which are not discernible to naked eyes. It had previously been unclear that topographic differences contributed to the variations in brightness, persistent over lunar cycles. With his telescope Galileo devoted substantial time to studying the moon in detail and intently, making many drawings of his views, which survive; many are printed in his small book *Starry Messenger.*

So, with this discovery by Galileo of a satellite system about Jupiter and the phases of Venus that everyone with a telescope could see, opposition to the heliocentric worldview vanished. Right? Wrong! Galileo became an unfortunate example of the reluctance of society to abandon views in which it has so much invested. Galileo was prosecuted for heresy by the Roman Catholic Inquisition and, after one day in prison, was confined to house arrest for the last decade or so of his life, because he refused to recant his writings and beliefs related to the heliocentric worldview. He was only officially exonerated by the church nearly four hundred years later.

Despite the unhappy ending, this story is altogether marvelous. The advent of an instrument with new capabilities most often leads to major scientific discoveries, the telescope here being a prime early example of the coupling between science and technology, as mentioned in the Introduction. In this case development of the elementary principles of optics led to the invention of the telescope, which in turn led to remarkable new science — dramatically increasing knowledge of how the world works.

The Speed of Light

The speed of light plays a key role in much of modern science, especially in astronomy. We move now to discuss the history of our knowledge of this speed and the major role of Galileo and serendipity in its determination.

In modern life, we are quite familiar with light traveling very fast. For example, we cannot discern any interval between our flipping on a light switch and our detecting the resultant light, for incandescent bulbs. (There are delays, unrelated to the speed of light, for fluorescent and LED bulbs.) The ancients were also quite familiar with this phenomenon, albeit not from flipping a light switch. In fact, Aristotle reputedly concluded that light travels instantaneously between places; that is, it travels infinitely fast. An early attempt to measure the speed of light was proposed nearly four hundred years ago by none other than our friend Galileo in his *Two New Sciences*. He suggested having two people, each outfitted with a lantern or a lamp, with one of them also equipped with a sand clock or equivalent. They could be separated by, say, about 10 kilometers, with each on a hill or mountain so that each could see the other, with the unaided eye (or perhaps with a small telescope).

The procedure was simple. Both lamps would be shielded from the distant person's view by the use of a piece of cloth, or equivalent, in his—there then being no "hers" in this field—hand. The first person, the one equipped with the clock, would draw away the cloth from in front of his lamp and, as simultaneously as feasible, start the clock; when the second person saw the light from the other's lamp, he would draw away his cloth from in front of his lamp, which would then be visible to the first person. When the first person saw the light from the second person's lamp, the first person would stop his clock. The experiment was doomed to dismal failure. Why? I mention here one of the many reasons: the delay caused by the uncertainty in human reaction time swamped the round-trip light travel time, by a factor of perhaps ten thousand or so!

Interestingly, our very same Galileo was instrumental in the first successful detection of the speed of light, albeit in a rather indirect way. We need to set the stage by describing a problem in navigation experienced by sailors of the time: How does one tell where one is when at sea, far from land, where land maps are not very useful as guides? It was the Age of Discovery, and long-distance trade, from the very late fifteenth century onward, was based mainly on oceangoing vessels that needed to navigate. Basically, to locate themselves on the earth's surface, sailors needed to know the values of two coordinates, traditionally latitude and longitude. How were they determined? Latitude is relatively easy to estimate at night by measuring the elevation angle of the North Star above the horizontal. This description applies for travel in the northern hemisphere; a similar setup works for travel in the southern hemisphere, where the Southern Cross holds sway. For daytime determinations, our sailor could wait until local noon and measure the angle the sun makes above the local horizon in the meridian containing the sun at that time. That measurement plus readily available information taken with him on the voyage about the sun's angle at noon with respect to the plane that contains the earth's equator could be used to determine the latitude of the place at which he was making the measurement. I omit the details and leave it to you to draw a figure to illustrate this determination, if you wish. Of course, either in daytime or at night, respectively, all bets are off for astronomical navigation if the weather makes sighting the sun or the stars infeasible.

What about the other coordinate, longitude? Suppose you know that your longitude on the earth is zero; that is, that you are on the Greenwich meridian, which passes through Greenwich in England. As the earth turns on its axis, different stars will pass over your meridian during a 24-hour period. At midnight, for example, one particular star will pass overhead. Similarly, other specific stars pass overhead, or cross your meridian, each at a specific time of day each day. So—and this is the key point—there is a tradeoff between time and longitude. If you are at sea and know the star overhead as well as the time, you can correctly infer your longitude. But any error in your clock will cause a corresponding error in your inference of longitude. Each hour of clock error causes a 15-degree error in longitude. Do you know why? Here's a hint: $360 = 24 \times 15$.

For sea voyages of months, with clocks of that period on ships being virtually useless over such lengthy time intervals, what could be done to solve this navigation problem? One way would be to travel directly to the latitude of your final destination and just continue along it until you reach the desired destination. But the trip would then most likely be considerably longer and be less efficient by also requiring more provisions and consequently having less room for the money-making cargo. Also, bad currents and bad weather would then not be easily avoidable by a change of course. It would in addition be difficult to recover from being badly blown off course. So people wanted good clocks.

What sort of astronomical clocks were available for navigation at the end of the fifteenth century? There was the lunar clock. However, the moon has a monthly period, rather long for determining the time of day. After his discovery of four satellites of Jupiter, Galileo suggested that the innermost one, Io, with an orbital period of only about forty-two hours, nearly twenty times smaller than that of the moon, might be a useful target. One could observe Io either when it entered into or emerged from Jupiter's shadow, thereby determining time to a much higher degree of accuracy. There were two problems with this idea: first, whereas in clear weather the moon is mostly bright and easily visible, either at night or in the daytime, Io is not. A telescope had to be used, and even so, Io was visible only at night; the telescope enlarged the image as well as, unfortunately, magnifying its apparent motion due to the ship's bouncing in the waves at sea. Finally, the problem was solved in the mid-1700s by John Harrison, who won the award set up in 1714 by the British government of £20,000 for the first person to solve this problem. He invented a mechanical clock, or chronometer, which in its fourth iteration could keep time to within five seconds over a month. (Collecting this reward turned out to be another matter—an interesting, if ugly, tale, which we skip.) This story exemplifies the unity of science: the attempt to use astronomy (exemplified in this case by Io's orbit), combined with engineering (the creation of a ship-proof clock), to solve a navigation problem, and the brilliant and unsuspected result from that attempt.

In order to use Io as a clock for navigation, scientists needed accurate predictions of its eclipse times: Io's disappearance into, and re-emersion from, Jupiter's shadow. To produce accurate predictions required making many measurements of Io's eclipses, extending over a long period of time. In the Paris Observatory, then under the direction of Italian-born Domenico Cassini, such a program of eclipse observations was undertaken in the late 1600s. An extremely intelligent Danish scientist, Ole Rømer, only in his midtwenties, was brought by Jean Picard from Denmark to Paris to help make these observations. Picard was visiting Denmark from the Paris Observatory on a related scientific mission—to accurately determine the relative longitude of the Paris Observatory and Tycho Brahe's observatory in Denmark. (The purpose of this determination was to use Tycho's superb planetary measurements in conjunction with those made at the Paris Observatory and elsewhere to improve planetary orbit determinations. Combining of all of those data made key use of accurate measurements of the relative locations of the observatories at which the sets of observations were made.)

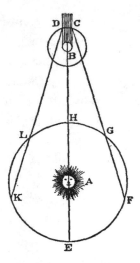

Figure 3.2. Sketch of the geometry of observations of Io's eclipses as seen from the earth at various places along the earth's orbit about the sun, all with respect to Jupiter (that is, Jupiter is kept at one position on the sketch so as to enhance its clarity, with the main point thereby easier to make clearly—or so I think). From Ole Rømer, "Demonstration touchant le mouvement de la lumière trouvé par M. Roemer de l'Academie Royale des Sciences," *Journal des Sçavans* 269 (1676): 233–236.

Ole Rømer took this assignment of measuring eclipse times of Io seriously; being very bright and observant, he noticed something unexpected in his data. Before describing what he noticed, let's first examine the geometry of Rømer's observations (fig. 3.2). Both the earth and Io move counterclockwise in their orbits about their parent bodies, the sun (Point A in the illustration) and Jupiter (Point B), respectively, if we look down on them from well above the earth's North Pole. Consider the earth at Point F in its orbit. From this point, as from any point at which the earth is to the right of the position in the figure of the sun-Jupiter line (line A–B in the illustration), Rømer could observe Io enter eclipse (Point C) but not emerge from eclipse (Point D): at the latter point, Jupiter would be in the way. Similarly, when the earth is on the left of the sun-Jupiter line, at Point K for example, Rømer could observe only Io's emergence from eclipse (Point D), but not entrance into eclipse (Point C). However, this distinction shouldn't affect the difference in successive times of entrance into, or successive times of emergence from, eclipse.

Examining his data, accumulated using the best clocks of his day, Rømer noticed something strange. He looked at the difference in times of entrance into eclipse of two eclipses separated by several dozen or so eclipses. He also looked at a corresponding difference in times of emer-

gence from eclipse. He noticed that the difference between the two entrances was less than the difference between the two emergences, by about
twenty minutes or so. What could possibly be going on, he must have asked
himself. Thinking out of the box, he probably said, "Maybe the answer is
a finite speed of light!" If light had a finite speed, then when the earth is
approaching Jupiter—that is, on the right side of the sun-Jupiter line—light
has a shorter distance to travel between successive eclipses than when the
earth is on the left side, when the distance between the earth and Jupiter
increases between successive eclipses. So for the times between successive
eclipses when the earth was approaching Jupiter, the elapsed time would
be less than for the corresponding situation when the earth was moving
away from Jupiter. For this difference in the times of emergences from, and
entrances into, eclipses to be all explained by a finite speed of light, then
that speed must be about one astronomical unit (approximately the sun-
earth distance) in about eleven minutes. In terms of our modern value for
this speed, Rømer's value was between two-thirds and three-fourths as large.
Not bad for a first try on such a difficult speed to discern at all! There is no
record of Rømer's having ever calculated the speed of light, per se, in earth
units of distance and time. That was apparently of no interest to him; his
main interest was apparently in showing that the speed of light was finite
and measurable, not infinite.

It is not clear to some scientists who have looked into the available evidence that time measurements were reliably accurate to a level that allowed
the determination of even close to that accuracy. But to my knowledge, a
comprehensive collection of the relevant observations no longer exists (an
early eighteenth-century fire in Copenhagen, to which Rømer had by then
returned, destroyed most of Rømer's data, I understand), and the drawing of
conclusions of this sort are therefore to be questioned, not accepted at face
value. However, Rømer's accurate prediction (see below) of an eclipse time
of Io—observably different from Cassini's prediction—would then have to
be considered mostly a mere coincidence. I confess to not being a great
believer in such coincidences.

Rømer had thus serendipitously made the first measurement of the speed
of light when trying to accumulate data on eclipses of Io for an entirely

different, navigational, purpose. As Louis Pasteur, nearly two centuries later, famously remarked: "Chance favors only the prepared mind"—an apt description of Rømer's.

Cassini, by the way, was rumored to have made the same deduction as had Rømer, but much earlier (see below), in August 1674. Then, purportedly, Cassini tested this deduction against corresponding measurements of eclipse times from the other of Jupiter's moons that had been discovered by Galileo. Measurements on the eclipses of these other moons supposedly did not support the deduction on the speed of light made from observations of the eclipses of Io. The data on these other moons did not show the same effect consistently, and therefore, it was supposedly said, Cassini at least at first rejected Rømer's claimed measurement of the speed of light. This description of history is now rejected by most, on the basis of a more reliable investigation. (History is a tricky thing to get "right," especially when, as is sometimes the case, the written evidence is either missing or ambiguous or some of those involved wished to hide certain actions.)

Previously, many people thought that the speed of light was infinite. Most famously, as already stated, Aristotle put forward such a belief, perhaps common in his time, nearly two thousand years earlier. Was Rømer's contrary result quickly accepted by other scientists? Pretty much! How come? Because Rømer had made, about two months earlier, a prediction that turned out to be correct of the time of the Io eclipse of November 9, 1676. His prediction was for a time ten minutes *later* than the corresponding prediction of his boss, Cassini, who had apparently claimed confidence in the accuracy of his own prediction, as it was the result from his latest method. At the November 26, 1676, meeting of the French Academy of Sciences, Rømer, not a member, gave a report on his results on the Io observations and his theory regarding the finite speed of light. Apparently the members in attendance were very impressed, given the accuracy of Rømer's prediction regarding Io's eclipse earlier that month. Rømer's presentation was published by the academy early in December 1676, with an English translation being published by the Royal Society in London seven months later.

I was curious about this possible contretemps between Cassini and Rømer, the former being, as noted above, the boss of the latter, and having been humbled by the latter in the eclipse-prediction face-off. I couldn't find much information on it but did learn that a Dutch amateur historian of science, Emile van Kreveld, was an expert on this subject. So I wrote to him and he answered. The upshot is that there is no evidence for any prior discovery by Cassini, but there is evidence that Rømer received a 60 percent increase in salary (or bonus?) right after his discovery, indicating that his accomplishment was indeed recognized and probably appreciated by Cassini, who as Rømer's boss presumably had some say over Rømer's salary.

Rømer accomplished much in his later life, too. Returning to Denmark in 1681, he was very much appreciated there. He was given a number of prestigious positions, some he held simultaneously: professor in Copenhagen University, astronomer royal of Denmark, master of the Mint, inspector of ships, chief magistrate of Copenhagen, prefect of police, mayor of Copenhagen, and member of the king's privy council. He also designed a unique new observatory, apparently made more planetary observations than even Tycho, designed new astronomical instruments, and invented a superior design for gears, a thermometer, and a micrometer. He probably slept little. He was a true polymath and, supposedly, a very nice person.

As technology improved, it became possible to measure the speed of light accurately in the laboratory. A bit after the mid-twentieth century, this speed, then and now universally labeled c, was known fairly well, to nine significant figures:

$$c = 299{,}792.458 \text{ kilometers per second.} \tag{3.1}$$

Since c is believed to be constant and universal, the above value of it is now used to implicitly define the unit of length, with the unit of time being defined by an atomic property. These units, agreed upon internationally in 1983, replaced the physical units previously adopted by the world. Thus, when I was young, the unit of length for the world was defined by a meter bar kept under controlled physical conditions in a vault in Paris. Now it is defined in part by the speed of light in vacuum. Times change.

FOUR

The Law of Gravitation, the Discovery of Planets, and Special and General Theories of Relativity

What happened next? The legendary Isaac Newton enters our center stage. After Kepler's major advance, some scientists later in the seventeenth century sought a more fundamental basis for a model of the motions of planets. Kepler himself talked about magnetism being—somehow—responsible, with its effect decreasing with distance from the magnet. Clearly, whatever force it was, it was not a contact force. (What *is* a contact force? It is a force in which the exerting entity is in direct contact with the object on which the force is exerted; it does not act at a distance, as does a force between the sun and a planet.)

Some people during this late seventeenth century, such as the well-known polymath Robert Hooke, thought that the force was likely of the inverse square law variety. What does this mean, and what might motivate its choice? In answer, let us give one example. Consider an ideal, perfectly spherical, balloon partially blown up. Its surface area, A, is proportional to the square of its radius, r, in accord with the basic geometric property of spheres. That is,

$$A \propto r^2, \tag{4.1}$$

where the middle symbol denotes "proportional to." What about the mass of the balloon? As we continue to blow up the balloon to larger and larger sizes, the mass will stay the same—except for the mass of the added air we blow in. But the mass per unit area of the balloon's surface will decrease because the area is increasing and we are neither creating nor destroying the balloon material, just changing its size. Since we have neither created nor destroyed any of its surface during this blowing up, the same mass of surface will cover a larger area, in proportion to the square of the balloon's radius, so as to conserve the amount of matter in the balloon material. We can represent this effect by a mathematical equation in which the subscript "o" represents the initial (that is, starting) values of the mass per unit surface area, m, and the radius, r, of the balloon in its partially blown-up state:

$$m(\text{unit area}) \times r^2 = m_o(\text{unit area}) \times r_o^2, \qquad (4.2)$$

or, dividing both sides by r^2,

$$m = m_o \times r_o^2/r^2, \qquad (4.3)$$

which, expressed in words, can be restated as: the dependence on its radius of the mass of the balloon per unit surface area (one unit of surface area) equals a constant, in this case written as $m_o \times r_o^2$, divided by the square of that radius. That is, *as the balloon's radius increases, the mass per unit surface area of its surface decreases in inverse proportion to the square of that radius.*

One might therefore consider a force that acts on materials depending on their distance from the source of that force to vary in a manner proportional to the inverse square of that distance. Why? In analogy with the way the mass per unit area of the surface of a balloon decreases as the balloon is blown up, one might think of the force as acting on the surface of the imaginary balloon and being spread out uniformly over its surface, whose area is proportional to the square of the radius of the balloon. If we assume that the force remains in a total sense constant as it acts on a larger surface, then one must consider that the force weakens with the inverse square of the balloon's radius. This weakening is necessary so that the total force can maintain itself constant. That is, r^2, the square of the radius of the balloon, which is proportional to its surface area, when multiplied by the force, which is proportional to r^{-2} (inverse square of the radius of the balloon), yields a result independent of radius: $r^2 \times r^{-2} = 1$. This concept of force is not an unreasonable one to test against nature's behavior.

Relative to its governing the motion of planets, we can describe this inverse square law of force as follows: each planet moves as if it were acted upon by a force proportional to the sun's mass and inversely proportional to the square of the distance between the planet and the sun (in other words, getting weaker the farther the planet is from the sun).

According to legend, such an idea was discussed at a dinner of intellectuals at a tavern near the Royal Society in London in the middle 1680s, following a meeting at the society. Supposedly in attendance were Edmund Halley (of comet fame), Christopher Wren (a highly accomplished mathematician and the architect of Saint Paul's Cathedral in London), and Robert Hooke (whose genius paled only in the shadow of Newton). Whereas an inverse square law sounded plausible, as just noted via a detailed analogy, what sort of motions of planets (orbits) would result from such a force law? Halley suggested that he seek Newton's opinion as the two were friends. When Halley later posed the question to Newton, he responded immediately: a planet's motion would be elliptical. Halley was apparently astonished. How could Newton answer so rapidly and so definitively? The answer was simple: Newton had calculated the consequences of his proposed inverse square law of gravitation about twenty years previously but had not gotten around to publishing it. Halley apparently urged Newton to swiftly publish his theory of gravitation. Two months later, Newton sent Halley a nine-page paper that showed how Kepler's laws followed from Newton's law of gravitation. And in 1687, after eighteen months of feverish work, Newton brought forth in print his now famous *Philosophiae naturalis principia mathematica* (usually just called the *Principia*), which covered far more than his theory of gravitation and, surprisingly, contained *no* equations. All of his arguments used only words and were based, where appropriate, on geometrical reasoning. I confess, however, to have never myself seen a copy of the *Principia*; all the information here is secondhand.

I have also heard a story that Newton's delay in publication was for a different reason. In this telling, Newton compared the acceleration of the moon in its orbit around the earth with that of an object dropped onto the surface of the earth (recall my fork experiment from the Introduction) to test their consistency with each other, given the proposed inverse square force law. To explain, albeit in oversimplified fashion: The moon is in orbit about the earth. It is falling—accelerating—toward the earth as it moves sideways, falling just enough so that it remains always at nearly the same distance from the earth. As a familiar analogy, consider that when you throw a ball parallel to the earth's surface, it falls as it moves sideways away from you. The faster you throw it, the farther it goes before falling back to the earth. A satellite goes so fast that it keeps on going, in orbit. The point here is that the acceleration, due to the earth, follows the same inverse square law for the moon as for the fork in the Introduction. The same earth acts on both.

Newton's calculations of these two accelerations disagreed with expectations by about 10 percent, an amount well outside his combined estimated errors for determining the acceleration of both bodies due to the earth. However, Jean Picard (see chapter 3) in 1669–1670 in France carefully re-determined the radius of the earth from surface measurements of the length of a degree of latitude and found it to be about 5 percent larger than the previously accepted value. When Newton years later learned of Picard's corrected value, he recalculated the acceleration at the earth's surface, which depends on the inverse square of the radius, and found it about 10 percent less than before. (Why did 5 percent become about 10 percent? The answer is because the acceleration depends on the *square* of the radius and $[1 + x]^2 = 1 + 2x + x^2$, which is about $1 + 2x$ when x is very small compared to 1, as in this case.) Newton then obtained agreement between his expectation for the moon's acceleration by the earth and that for an object near the surface of the earth, given his proposed inverse square force law of gravitation. He then proceeded with the preparation and publication of the *Principia*.

Of relevance to our present discussion, the *Principia* contained not only Newton's proposed law of gravitation but also, among other scientific work, his corresponding laws of motion. We will shortcut the discussion and describe only two of these. The first is the famous

$$F = ma, \tag{4.4}$$

where F is the force acting on the body of mass, m, which as a result undergoes an acceleration, a. (Here a is *not* the semimajor axis of a planetary orbit. This latter quantity and acceleration are both abbreviated by virtually all scientists by the same symbol. Context should allow you to always distinguish which of the two quantities is being abbreviated by this symbol.) This equation, 4.4, represents Newton's well-known relation stating that the acceleration of an object is proportional to the force acting on it with the constant that describes the proportionality being the object's mass, as represented succinctly in symbols through the above equation. This equation appears to be very simple. But, in fact, those few symbols hide a plethora of subtleties that appear in almost any real application of it to a physical situation in nature. There is also the basic question: What is the mass of an object? Let us skip a discussion of the answer, save for noting that mass is how much stuff an object contains and, operationally, is the quantity in equation 4.4 that allows it to work correctly!

On to the second law, the inverse square law of gravitation:

$$F = GMm/r^2, \tag{4.5}$$

where F is the gravitational force between two pointlike (see below) objects of masses, M and m, with r the distance between them; G is a constant of nature, now universally called Newton's constant of gravitation. In words, the equation states that two point masses (we mean, by this expression, two masses, each sufficiently small so as to be of negligible size compared to that of everything else around) attract each other with a force proportional to the product of these masses and inversely proportional to the square of the distance between them. Here the force is definitely not a contact one but action at a distance. This means, in this context, that the sun exerts a force on a planet from a distance. Notice, too, the important fact that the planets also exert a force of this same type on each other and on the sun.

The mathematical problems posed by this reciprocal aspect of the simple-sounding law are exceedingly complicated; they stimulated rather impressive research in applied mathematics for several centuries following Newton's publication of his laws, another example of the unity of science — if one allows mathematics under the tent. Finally, note that this latter, reciprocal, aspect of Newton's law of gravitation is completely missing from Kepler's laws. How is it, then, that Kepler's laws could predict so well the observations of planetary positions? The sun's mass is so much greater than that of all of the planets that a good approximation to planetary orbits can be obtained by considering only the effects of the sun on each planet. For example, the planet with the largest mass, by far, is Jupiter, but it has a mass only one-thousandth that of the sun. Although in Kepler's day the effects of the planets on each other's orbits were not easily discernible, they are now dramatically larger than the far, far smaller present measurement uncertainties. These deviations thus are now noticeable in a short time, and have been now for over two centuries, as we will discuss below.

Newton's laws have another remarkable property exhibited when we combine them, by setting the right sides of equations 4.4 and 4.5 equal to each other, since both are equal to the F (force):

$$ma = GMm/r^2. \tag{4.6}$$

The m on the left seems equal to the m on the right, allowing us to cancel one out against the other, leaving a remarkable result: in Newton's laws of motion and gravitation, the gravitational acceleration (a) of a body, relative to another, is *independent* of its own mass. Thus, for example, two different masses will fall at the same rate on the—much larger—earth. Truly remarkable!

The principle here, called the Principle of Equivalence, and placed on a pedestal by Einstein, is often discussed in terms of an apparently apocryphal experiment, purportedly carried out by Galileo at the Leaning Tower of Pisa. He supposedly dropped simultaneously from the top of the tower two objects of different masses, noting that they descended at the same rate, thus also hitting the ground below simultaneously, as we demonstrated in a more modest setting in the Introduction. The Apollo astronauts demonstrated similarly on the moon by dropping a feather and a wrench, which reached the surface simultaneously. This amazing property of nature has been verified with increasingly higher precision over the past 1,500 years or so, starting long before Galileo. The most recent results show that this principle holds to at least within a few parts in a trillion! Why should anyone try to make more accurate measurements? We are prompted to probe deeper because this Principle of Equivalence, like all our models, may break down at some higher level of accuracy and disclose another layer of nature's behavior. In fact, later in this chapter, we will encounter a prime example of this type, which led to a model of nature's behavior that succeeded Newton's.

How may we proceed to calculate the orbital motions of the planets from Newton's laws? The mathematics required is far beyond the level used here. I wish, however, to point out one other potentially major issue: Newton's law speaks of each mass point attracting each other mass point. . . . But are the planets mass points? Is the sun? No, you may conclude, not by any reasonable definition of a point. They are (nearly) spheres. One could, of course, argue that compared to the distances between planets, and between planets and the sun, the size of a planet is negligible, so that one could ignore any effect of planet sizes in calculating planetary orbits about the sun. And that is true enough at our present measurement accuracies. But with the earth-moon system, the ratio of the earth's radius to the earth-moon distance is only about 1:60, which is not at all negligible if one wants to do accurate (errors well under 1 percent) comparisons between observations and computations.

From the spherical symmetry of a uniform sphere, one might guess that the gravitational force, under Newton's laws, might act as if all the mass were concentrated at the sphere's center. This guess is correct. But proving this conjecture with the mathematical tools then available was far from easy. Newton, however, was a superb geometer, on the same plane as Euclid (pun intended), and was able to prove what was then a remarkable theorem: for a spherically symmetric distribution of mass, the gravitational force exerted by it on external mass points was the same as if all the mass of the spherically symmetric body were concentrated at its center—that is, as if it were a point mass. An impressive result, which seemed to solve a major problem.

You might well wonder, however: Is a planet spherically symmetric in its mass distribution? No. But deviations from this condition are fractionally small; because they are so small, they can be handled in a simple systematic way, which is nonetheless beyond the scope of this book. With this result of Newton's in hand, the way was cleared to make accurate, albeit difficult, calculations of lunar, planetary, and satellite orbits, consistent with the values of the corresponding observations, all made from the earth. This calculational task was taken up by Newton along with many contemporary and following astronomers and mathematicians, leading to major advances, not only in our understanding of solar-system bodies' motions, but also in applied mathematics.

Wait a second, you might well say: How could these scientists have made any progress in calculating these complicated effects of the forces the planets exerted on each other? Wouldn't the scientists have had to know the masses of the planets, where the planets were all located at a given time, and so on? Yes! These values are critical and were certainly not available from the start. What could one do? The simple answer is that the theory could be developed initially without specific values for these unknowns, which could be determined through comparison of the theory with the observations. Consider a simple analogy—the straight-line relation between distance and time traveled by a car motoring on a perfectly straight road at a constant speed. Initially, we do not know the car's speed. From a few measurements of the location of the car, relative to its starting point, and the time at which the car reached each of these locations, we can infer its average speed. Thus, from these measurements, we can determine the value of the unknown speed. Similarly, but in a far more complicated manner, we can use many observations of the positions of the planets with respect to the fixed stars, and the corresponding times of these observations, to determine the orbits and the masses of the planets. This situation is very similar to Ptolemy's where he needed the values of seven quantities to determine the orbit of a planet for his model of planetary motion.

The Discovery of New Planets

Out of the blue so to say, in March 1781, the now and even then famous astronomer William Herschel made a startling discovery: the existence of a planet, not visible to the naked eye, farther from the sun than Saturn. This was the first new planet to be discovered since antiquity. At first, Herschel thought that he had discovered a new comet. It took some months before scientists understood that it was a planet. Being a newly discovered object, it generated much interest and was observed quite often. From these observations, its orbit was determined, showing it to be on a roughly circular path, well beyond Saturn's orbit. Most comets, by comparison, had rather elongated orbits that approach close to the sun. This, in addition to the fact that this object didn't seem to have a tail, gave it the "smell" of a planet rather than of a comet.

Given the state of knowledge and the reach of science, this discovery of another planet, soon to be named Uranus, was indeed big news. Even Benjamin Franklin wrote, albeit some years later, to congratulate Herschel. The story of the naming of Uranus is interesting in its own right; the curious can find this story on the Web.

There had been some prediscovery observations of this newly discovered planet, as far back as the 1690s, but they were not recognized then as being of a previously unknown planet. They had been recorded as being observations of stars and neither looked at nor thought about again until after the discovery of Uranus. Recorded in observation ledgers at those earlier times were the epochs (that is, the times) of the observations, and the sky coordinates of the objects. There was no photography, no film, nor, of course, any electronic or digital recording devices. Recall, once one has observed a fair fraction of a planet's orbit, one can project with reasonable accuracy where it had been in the sky at previous epochs, as well as where to look for it in the future. So, if one has good eyesight, patience, access to observer ledgers, and the fortitude and interest to search, one can go through and check entries that recorded the predicted positions in the sky where that object was calculated to have been at different epochs.

Why is it of any interest to find such earlier sightings? It is true that the longer the time span over which observations of a planetary body extend, the more accurately one can determine its position in the sky. By incorporating observations of what they now knew as Uranus from the past century, astronomers were thereby able to increase the accuracy of their estimation of the orbital period of Uranus about threefold. As time goes on, will the importance of such prediscovery observations become more or less important? Why? As time goes on, and new, more accurate (why?) observations accumulate, the old ones will be less important for determining the object's orbit, but still useful.

There is a spectacular detective story lurking here, filled with many intriguing twists. We thus skip now to 1820. By this time, astronomers had been assiduously observing for almost half a century the newly recognized member of the planetary system. After all, Uranus was the new kid on the block and hence drew much attention. In 1820, astronomers noticed that the predicted orbit of Uranus, obtained by extrapolating from past observations, did not match the new observations within their expected uncertainties. The differences were systematic—that is, they had a definite trend; they weren't just randomly scattered about the predicted orbit. Many astronomers explained away these differences as not being large enough to worry about. But by the early 1840s, the discrepancies were glaringly large, and no sensible astronomer could then ignore them.

What could be going on? Were, for example, Newton's laws not up to the task? Had the laws reached their limit in accounting for nature's behavior? Independently, and unknown to each other, John Couch Adams in England and Urbain J. J. Le Verrier in France had the same idea. What if there were yet another planet beyond Uranus's distance from the sun, massive enough to noticeably distort Uranus's orbit? How could we check on this possibility? It was hard enough to calculate the effects of planet A on planet B knowing planet A's mass and orbit fairly well. But how could one possibly do such a calculation without such information?

Though these two scientists used different techniques, both needed to make *assumptions* to answer this question. Most important were assumptions about the orbit. All planets known at that time had orbits that lay nearly in the same plane and had, except for Mercury, nearly circular orbits. So Adams and Le Verrier independently made two corresponding assumptions: the orbit of this putative planet lies in essentially the same plane as does the earth, and its orbit is circular. These two assumptions, however, were not enough. Both astronomers independently went to an empirical law of sorts that had already been formulated, which they thought would give them a way to determine a reasonable value for the semimajor axis, *a*, of the orbit of this putative new planet.

This empirical law was called Bode's law, or the Titius-Bode law, after the two people associated with it (check the Web if you'd like to know more). Because this law had predicted Uranus's semimajor axis to within about 2 percent accuracy, both Adams and Le Verrier thought that it could also be used to determine the putative new planet's semimajor axis. In fact, they were quite wrong, but luckily this error turned out not to be crucial because of the location in the sky of the new planet at the time of its discovery.

With these assumptions, somewhat different mathematical procedures, and well over a year's hard work—there were no computers then, mainly just tables of logarithms and hard work—each of the two astronomers arrived, as his end result, at the direction in the sky in which this putative planet would be found at that time and with approximately what brightness it would shine. Hence, interested astronomers could determine whether their instruments were sensitive enough to see such a starlike object, and if so, they should look.

With his prediction in hand, Le Verrier tried to convince French astronomers to search for this planet, but none would do so on the mere say-so of a theoretician. In England, the story was somewhat different. At the end of the summer of 1845, Adams had sent his preliminary location of a possible new planet to James Challis at the Cambridge Observatory. In July 1846, Challis began mapping the stars in the vicinity of the location Adams had provided. But Challis didn't follow up to see what he might have discovered by looking at his observations; he was occupied then full time trying to put accumulated observations of comets in order. In fact, Challis had recorded seeing this new planet twice but did not realize it until after the putative planet had been identified. Adams, not knowing what Challis had done, continued to go around in the summer of 1846 trying to seek observations, even going to the home of the astronomer royal and knocking on his back door. He left a message, which apparently was delivered: the astronomer royal did foster some observations, but apparently no one took the time to check what, if anything, had been discovered.

Frustrated with his own astronomy colleagues, Le Verrier turned to Germany. At the Berlin Observatory was an astronomer, Johann Galle, who had sent Le Verrier a copy of his doctoral thesis the year before for comment. Le Verrier thus wrote Galle on September 18, 1846, urging that he look for this predicted new planet, telling him where to look and what he might expect to see. Le Verrier estimated that his predicted location would be within about a degree of the actual location and that his predicted brightness would be within about a factor of two of being correct—that is, the observed brightness would be within half and twice the value predicted. Galle received this request on September 23. After some effort, he obtained permission from the observatory director, Johann Encke, of comet fame, to make this observation. Galle instructed his eager assistant, Heinrich Louis d'Arrest, to look that very night for a new object, using their top-of-the-line telescope: a 23-centimeter-diameter refractor.

D'Arrest saw the object at nearly the predicted place and at nearly the predicted brightness. The values of latitude encompass −12 to −15 degrees; the difference between the observed and calculated latitudes is thus far less than 1 degree (fig. 4.1). Remember that 1 hour of longitude is 15 degrees; 4 minutes of longitude (1/15 hour) thus equals 1 degree, which is about the difference between the observed and calculated longitudes.

This discovery made Galle instantly famous and added to Encke's fame. D'Arrest was not so fortunate, being too low on the totem pole to gain much recognition. The discovery itself was truly world news: a new planet had been discovered at almost the exact spot in the sky that it had been predicted to be! Then, after this discovery was publicly announced, the British looked at their earlier data and also found the new planet. (Challis, for one, was

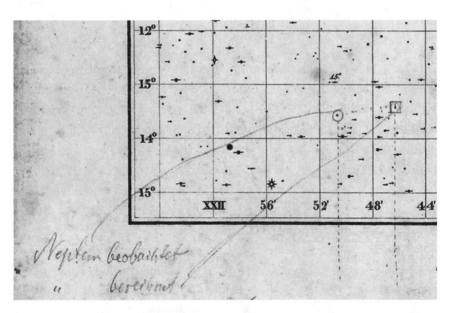

Figure 4.1. A comparison of the observed and predicted positions of Neptune. Note that *beobachtet* in German means "observed" in English, and, similarly, *berechnet* means "calculated."

Figure 4.2. A cartoon in *L'Illustration*, a French publication, from November 7, 1846, showing John Couch Adams searching without success for Neptune (left panel), until he spied it in Urbain J. J. Le Verrier's notebook (right panel).

mortified.) Thus started the real fun—the fight for priority. Some, but not all, of the relevant English claimed that Adams had predicted the position of this new planet first and demanded credit. The French felt, to the contrary, that the English were trying to steal their clear priority (fig. 4.2).

Quite a controversy ensued, and when it finally died down, the French had agreed to share credit for the discovery with the English, fifty-fifty. The English for their part bestowed on Le Verrier, but not on Adams, a prestigious award. One can only imagine how this history might have differed had Le Verrier ignored Galle's request for comments on his (Galle's) thesis, or had Challis looked at his results at the time of his observations, or . . . . Interestingly, we now know, but only for the past few decades since the discovery of this fact, that Neptune was observed more than two hundred years earlier by none other than Galileo; he almost recognized it as definitely moving with respect to the star background, but it moved out of his field of view while he concentrated on observing Jupiter.

Then came the battle of the name. As with the naming of Uranus, this is a story in and of itself (if you are curious, there is always Wikipedia!). Also, you might think that you could reliably infer that the notes in German in the lower left part of fig. 4.1 were made well after the discovery of Neptune. Why is that?

This saga illustrates the profound difference between a *prediction* of a new discovery and the *fact* of a new discovery. Neither English nor French astronomers could see the importance of trying to make such a discovery when based only on its prediction. Once made, however, the discovery fascinated the world. Aside from the sheer impressiveness of there being another planet, Neptune was the first one predicted and subsequently discovered; it thereby gave an impressive boost to science and scientists. Newton's invented laws were so good that they allowed prediction of a previously unseen planet. It was a dramatic development for those days and a tremendous feather in scientists' caps.

Astronomers, applied mathematicians, and physicists—or at least a subset of them—were flush with the thrill of this example of the power of science, clear to everyone, not only to the highly educated, but to the least educated as well. Did science then immediately sweep over the world and stimulate everyone to learn and contribute? While I know of no study that examined this question, I am personally unaware of any noticeable change in the pace of scientific development or in the funding of science following this discovery of Neptune.

The Advance of the Perihelion of Mercury's Orbit

In 1859, barely over a decade after his successful prediction of the existence, sky location, and apparent brightness of the planet Neptune, Urbain J. J. Le Verrier published the results of his study of the observations of the transits of the sun by the innermost planet, Mercury. His conclusion was startling: Mercury's orbit is inconsistent with the predictions from Newton's laws. This result had a familiar ring. Previously it had been Uranus whose orbit disagreed with predictions. Then it was proposed that Newton's theory might not quite work all the way out there. As we now know, it turned out that Newton's theory worked just fine and that the cause was a previously undiscovered planet. Would the same result hold here?

Before pursuing that question, we describe what a "transit" is in this context. A transit consists of a planet, in this case Mercury, passing in front of the sun as seen from the earth; the planet is observable from the earth as a black dot moving across the face of the sun. Very precise measurements can be made of the times at which the dot enters and exits the sun's disk, which in turn allow us to make accurate constraints on the orbit of the transiting

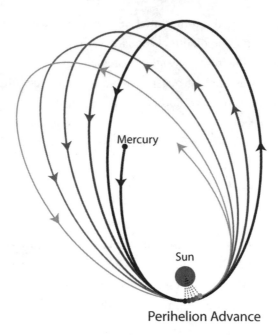

Perihelion Advance

Figure 4.3. An exaggerated picture of the advance with time of the perihelion position of Mercury's orbit. Courtesy of David Shapiro.

planet. Le Verrier examined the data on about twenty transits of Mercury, gathered from 1687 forward, the better part of two hundred years. He compared results for Mercury's orbit determined from these observations with theoretical predictions of Mercury's orbit, based on Newton's theory, and the orbits for the other planets, and he found a discrepancy. Specifically, he found that the perihelion position ("perihelion" means the closest [*peri*] point to the sun [*helion*] in the object's orbit) of Mercury's elliptical orbit was advancing more rapidly than predicted. With each successive orbit, Mercury's perihelion moves slightly (advances), in the same direction as the motion of the planet (fig. 4.3).

This advance was found to be moving about 40 arc seconds per century faster than his calculations had predicted. From the equations of Newton's models of motion and gravitation, eighteenth- and nineteenth-century scientists calculated that the gravitational effects of other planets, especially the most massive by far—Jupiter—and the nearest—Venus—caused Mercury's perihelion position to advance about 500 arc seconds per century. Yet

Le Verrier's analysis of the transit observations indicated a rate of increase almost 10 percent larger, in fact, about 540 arc seconds per century.

What could be causing this discrepancy between observation and theory? Scientists quickly devised four main proposals. The first was by Le Verrier himself, who suggested an unknown planet closer to the sun than Mercury. Just as the prediction and discovery of a planet farther from the sun than Uranus explained the apparent discrepancy in Uranus's orbit, so might a planet closer to the sun than Mercury explain this new discrepancy. This putative planet, dubbed Vulcan because it would be rather close to the sun and therefore very hot, was sought but never found. Within a few years or so, though, it was shown that the upper limit from observations on Vulcan's size, for reasonable values of its mass per unit volume (density), would leave it with too low a mass to be responsible for this discrepancy in Mercury's orbit.

Another suggestion involved a belt of individually small objects that or-bited the sun inside Mercury's orbit, which could contribute to an advance of Mercury's perihelion position. Ultimately, a combination of sensitive ob-servations and relevant calculations showed that such a belt did not provide a viable solution for this discrepancy in Mercury's orbit.

A further idea was that the sun, due to its rotation, was oblate, bulging at its equator like the earth, and that this bulge contributed to an advance of the perihelion position. If this bulge were large enough, its gravitational effect on Mercury could cause a sufficiently large extra rate of advance of Mercury's perihelion position to explain the discrepancy. Alas, the best ob-servations and calculations showed that this explanation, too, didn't work: the sun's bulge was far too small.

As a last resort: perhaps Newton's laws weren't quite right but differed from a more correct model in a small way that could be responsible for the orbital discrepancy disclosed by Mercury. For example, maybe the inverse square law was not quite correct and the exponent was slightly different from two. This type of ad hoc fiddling with Newton's laws didn't work either. There always seemed to be other differences between the observations and the theoretical predictions that prevented the proposed change from yield-ing agreement with all of the solar system observations.

What if Le Verrier had erred in his analysis? This possibility was largely eliminated toward the end of the nineteenth century by a reanalysis of the data, including forty years of more accurate data, by Simon Newcomb, director of the U.S. Naval Observatory. He obtained a discrepancy of 43 arc seconds per century, nearly the same discrepancy as Le Verrier had found, but with a smaller uncertainty, due to his having more, and more accurate, data to analyze.

This problem with the disagreement between theory and observation persisted for more than half a century. How was it finally resolved? It required a major revolution in our model of the theory of gravitation.

The Special Theory of Relativity

Enter Einstein. At the beginning of the twentieth century, he looked at the fundamentals of the theory of space and time and the theories of electricity and magnetism (E&M). The first was developed primarily by Isaac Newton, and the latter two developed primarily by Michael Faraday, via experiment, and James Clerk Maxwell, via theory, in the middle of the nineteenth century. Einstein then created a bold new model of nature's behavior—a new theory of space and time, the special theory of relativity. As was his wont, he took very simple, but often startling, starting points; in this case he took the equivalent of the following two concepts:

1. The speed of light, c, in vacuum is *independent* of the speed of the source of the light or of the receiver of it. This characteristic imputed to light by Einstein, and consistent with Maxwell's theory of E&M, is quite different from our ordinary experience with, for example, throwing balls. If we throw one from a car in the direction of motion of the car, the speed with which we throw it must be *added* to the speed of the car to match the speed of the ball measured by someone catching it in front of the car. Disregard here possible catcher-car collisions.

2. Nothing can travel in a vacuum at a speed greater than the speed of light. This limit is also consistent with Maxwell's E&M theory.

One question arises: Are the two above points independent of each other, or are they inextricably connected? Suppose that we postulated the second but not the first. What situation could we find ourselves in? We could launch a light signal from a moving platform and have the speed of the platform added to the speed of the light signal. But then we would have violated the second basic principle or at least created for ourselves an awkward situation with self-consistency seeming to be lacking. Other situations have been considered, too, all leading to the not unreasonable conclusion that these two points of special relativity go hand in hand.

Starting from these seemingly simple assumptions, Einstein turned our conception of space and time upside down (or inside out). They were no longer to be considered separate and distinct but inextricably mixed, somewhat like the two points above.

All sorts of odd results were predicted, based on Einstein's theory, and those susceptible to test have been verified. But these differences from our ordinary experiences are appreciable only at relative speeds reasonably near c. Thus, if an object has a lifetime well measured in the laboratory but is traveling by us at nearly the speed of light, we will find it to live substantially longer than the lifetime we measured for an identical type of object while it was at rest with respect to us.

The elegance of the logic of the special theory of relativity won the approval of many of Europe's top physicists almost immediately. It was the first step on the road to explaining the anomalous perihelion advance of Mercury's orbit, although it was certainly not recognized as such at the time. You may have noticed in this first step that we uttered not one word about mass and its behavior with other masses. Einstein was of course well aware of that limitation—a main reason for the modifier "special"—and set out to eliminate it: he needed to generalize his special theory to include masses and their interactions; that is, he needed to include a model for gravitation.

The General Theory of Relativity

First, we ask, what led Einstein to think that Newton's universal law of gravitation needed changing? To answer, let's review and analyze the statement of Newton's theory: every mass attracts every other mass with a force proportional to the product of the two masses and inversely proportional to the square of the distance between them (equation 4.5). What implicit property of this law is inherent in this statement? It is action at a distance. That is, the force one body exerts on another depends only on where the body is at the time (instant) it exerts its force; no time is allowed for the force exerted by the distant body to reach the body on which it is acting. As one body moves, its changed effect, according to Newton's theory, on another body will be felt instantly, no matter how far apart the two bodies are.

But if nothing can travel faster than c, then such instantaneous action at a distance violates Einstein's second postulate in his special theory of relativity. What to do? Einstein sought a new theory of gravitation that would not have this defect of action at a distance. Was it a simple matter to devise such a theory? Not exactly; it took Einstein almost a decade of rather intensive work to develop such a theory to satisfy this constraint and others. The theory was based on the previously mentioned principle of equivalence: in effect, one cannot distinguish the gravitational attraction exerted by one body on another from the corresponding acceleration of the other body, with the first body missing. The resultant intellectual creation, which Einstein called his general theory of relativity, was completed at the end of 1915, in the midst of World War I. It was a dazzling theoretical accomplishment. Word of it spread, even in wartime, through a network that connected scientists via neutral countries.

We expressed Newton's theory of gravitation in words and with relatively simple formulas. What about Einstein's theory of general relativity? It is a quite different animal, far more complex, and cannot be easily reduced to a few words that would mean much to the average reader. I will thus say only that the theory is expressed in a set of interconnected equations with the collection of left-hand sides representing the geometry of space-time and the right-hand sides expressing the mass, energy, and momentum contents of the universe. (For our purpose here, you can consider the momentum of an object as being the product of its mass and its velocity.) Thus, according to these equations, the mass, energy, and momentum in the universe determine its space-time structure. For these equations, only in a very few, rather simple cases are solutions known relevant to physical situations, although use of numerical methods now allows far more complex situations to be solvable.

Why have I presented even this brief view of the theory, given that its mathematics is way, way beyond the level used in this book? Because I think it is good for your soul to glimpse how much more complicated the fundamental theory of gravitation became between Newton's and Einstein's formulations. It is also a teaser for those who might wish someday to learn the details. Nature is obviously much more subtle than we had long ago imagined, and rather than getting simpler, our model of gravitation has become much more complicated. In one way, though, it became simpler: Newton needed both his law of motion as well as his (separate) law of gravitation; in Einstein's theory, both are contained in a single set of, albeit far more complicated, equations.

Let us return now to our discussion of Einstein and his creation of the theory of general relativity. When he had first put his new theory into acceptable form, what did he do? He calculated the theory's prediction for the orbit of Mercury about the sun and found, surely to his delight, that rather than a fixed ellipse as for Newton's theory for a single planet, the theory of general relativity predicted that the perihelion position for Mercury would advance by 43 arc seconds per century—exactly as inferred by Newcomb for the discrepancy with Newton's theory. Imagine Einstein's feeling to obtain excellent agreement between his new theory, on which he had worked for nearly a decade, and the anomalous observation that had remained enigmatic for over half a century!

What about the orbits of the other planets? Why didn't they, like Mercury, disclose a difference in their perihelion advances from the predictions of their orbits based on Newton's theory? In brief, this general relativity effect on the other planets was too small to then be observed. The effect was predicted to be larger as the semimajor axis got smaller and, of course, to depend on the accuracy with which one could measure the position of perihelion. The smaller the eccentricity, the closer to circular the orbit, and the harder it is to pick out the perihelion position. Since, for example, Venus's orbital eccentricity is about fifty times smaller than the eccentricity for Mercury's orbit, there was no chance for Venus's general relativistic perihelion advance to be observed in 1915.

What other testable predictions did Einstein's theory make different from those of Newton? Einstein realized that there were two others: first, that the frequency of spectral lines (see chapter 5) would shift when the source of the lines was near a massive object but the detector was far away; and second, that the direction of travel of light waves would shift as they pass by a massive object. Both were eventually measured, the second having been tested most accurately by a method that I suggested: use of VLBI (see chapter 10). Earlier confirmation of this second prediction, made from optical observations taken during a total eclipse of the sun in May 1919, led directly to Einstein's becoming and remaining a world-famous scientist, if not a household synonym for genius. When the German Einstein's prediction of the bending by the sun in the direction of starlight was verified by the English, this result was prime newspaper material and caught the attention of the post–World War I public.

I pointed out another prediction made by his theory, not envisioned by Einstein (after reading the remainder of this paragraph, can you think why?), that I tested in concert with many superbly talented colleagues. This prediction is that light would take longer to make a round trip to a target if the path of the light passed near a massive object. This effect is minuscule in the solar system. For example, the round-trip delay of the echo of a radar signal (see chapter 5) would be increased by about 200 microseconds (millionths of a second) were the signal to pass near the edge of the sun on its way from the earth to Mercury when Mercury was behind the sun as seen from the earth. The prediction of this effect was first tested successfully with interplanetary radar measurements, as noted above, and has since been tested far more accurately with interplanetary spacecraft that passed behind the sun. Observations and theoretical calculations of this effect—known as the Shapiro delay—agree to within better than one part in ten thousand.

This is another example of the unity of science: how, in astronomy, we can test a proposed fundamental law of physics, as was the case with Newton's laws before it. It also shows how scientific study can have amazing and unexpected results. Observations of an object in our outer solar system, and the deviation of these observations from predictions, led directly to the discovery of a new planet. Even more powerfully, a similar discovery in the inner solar system led in part to the discovery of an entirely new approach to the theory of gravitation that now also forms the basis of our view of the structure of the universe (see chapter 7).

In summary, let us note that observations of solar system objects, in quick succession—in an historical sense—took us from (1) deviations from predictions of locations of planets in the outer solar system that led directly to the discovery of a new planet, thus demonstrating dramatically the power of science, to (2) a similar discovery in the inner solar system that led in part to the discovery of an entirely new approach to the theory of gravitation that now forms the basis of our view of the structure of the universe.

This is the end of the line for our treatment of fundamental theoretical physics, until near the end of this first part of the book, when another problem that we will encounter may stimulate a further, deeper understanding that may unite the physics of the large with that of the small—an enigmatic comment now, but it should mean more to you after you read chapter 7.

FIVE

The Cosmic Distance Ladder

Now we change topics in our unveiling of the universe to a more practical one that is of key importance: the issue of distance. We need to know distances, for example, to develop accurate models of the universe. Thus, we want to know how far away from us are the moon, the sun, the planets, the stars, and the galaxies? Obviously, they are very far away, but *how* far? Because of the enormity of these distances compared to any we encounter on earth, getting handles on them has been a vexing problem since the beginning of human questioning about the universe. As it turns out, different tools are used for different scales of distances. Astronomers, with their wry sense of humor, dubbed the totality of these tools the cosmic distance ladder (CDL). Each tool, or distance range for which it can be used, is considered as a rung on this (figurative) ladder. The first, or bottom, rung applies to the distance scale within the solar system, to which we now turn.

Before we begin our climb of the distance ladder, though, a warning: part of this chapter is probably the hardest in the book. If you can understand it, the rest of the book (with the possible exception of chapter 11) should, I hope, be like a piece of good chocolate cake.

Solar System Distances

Because of its relative closeness to the earth, we can tell approximately how far the moon is from us via parallax, or triangulation (see the next section):

about sixty earth radii. What about the sun and planets? There we have an aid in Kepler's third law. How does it help? It relates orbital period to orbital semimajor axis, with orbital periods having been the easiest to measure accurately. And semimajor axes are expressed in astronomical units, with one astronomical unit being approximately the average distance between the sun and the earth. This unit of distance, as its name implies, is fine for astronomy. It is not so fine for sending spacecraft through the solar system. For that purpose, we need to know distances in earth (terrestrial) units. We know the power of rockets in terrestrial units of distance, for example, in metric units—centimeters, meters, kilometers, among other such terrestrially used quantities. What are the relations of these distance units to astronomical units? This is not easy to determine. Triangulating to a relatively nearby asteroid was about the best people could do before the last half of the twentieth century, and not very well at that.

By the mid-twentieth century, astronomers thought that they knew this relation with an uncertainty of about one part in ten thousand (10^4). However, when we were able to determine it more accurately (see below), it turned out that they knew it to only one part in one thousand (10^3), ten times worse than they had thought. Let's examine an effect of this uncertainty: suppose we want to steer a spacecraft to Venus but know the distance only to one part in one thousand, what sort of error would we make? A simple calculation provides a crude answer: (1.5×10^8, the approximate number of kilometers in 1 astronomical unit) × (1×10^{-3}, the fractional error of one part in a thousand) or, approximately, 1.5×10^5 kilometers. Such a spacecraft to go to Venus had to travel only about a third of an astronomical unit. Suppose this spacecraft were designed to include a camera to take close-up pictures of the planet; then, a miss distance of order $(1/3) \times 1.5 \times 10^5$ (= 50,000) kilometers could be quite serious. (The true uncertainty in the miss distance could not be determined this simply, but for our present purpose it is a satisfactory approximation.) For other instruments on the spacecraft, such a miss distance could also be serious. This was the situation destined for Mariner 2, the first spacecraft we were to send to visit another planet, in 1962. Different solutions to this navigation problem were possible in principle. What, in fact, did we do?

We used radar to solve the problem; this solution then became the first rung of the cosmic distance ladder. What is "radar"? Although now a common word, it is an acronym signifying *ra*dio *d*etection *a*nd *r*anging. During the Battle of Britain in World War II, radar helped to defend the British Isles against attacks by German warplanes: radio signals were transmitted into the sky, and the detected echoes allowed those warplanes' locations to be determined and suitable defenses to be employed (well told in the BBC drama *Castles in the Sky*). Radar, although invented somewhat earlier, was developed rather assiduously in World War II in all its aspects, including transmitters and sensitive receivers. Thus, in 1946, just after the end of the conflict, scientists both in Hungary and in the U.S. Signal Corps in New Jersey, apparently unbeknown to each other, were able to send radio signals to the moon and detect their echoes.

Since the moon and the sun appear to be about the same size in the sky, surely it should have been about equally easy to obtain radar echoes from the sun, right? Wrong—for two reasons: first, the radar echo power is inversely proportional to the distance to the target raised to some power (exponent). What power? The answer is a bit subtle to uncover and is a key to the difference between our detecting radar echoes from a body the size and distance of the sun compared to one the size and distance of the moon. As a radar signal moves out into space, it essentially spreads out on the surface of (a part of) a sphere, centered on the radar. The area of this surface increases with the square of the sphere's radius (a geometric fact; see the similar discussion in chapter 4 on the inverse square law). Hence, the power per unit area of the signal reaching the target has decreased with one over the square of the distance from the transmitter on the earth to the target: r^{-2}, where r is the distance from the transmitter to the target. The target will reflect some of the radar signal back to the transmitter. On this return trip the reflected radar signal also spreads over the area of (a part of) the surface of a sphere. As it travels further from the target this return signal (echo), too, also decreases in strength per unit area with the inverse square of the distance: r^{-2}. Altogether, then, we lose power by a factor proportional to the inverse of the distance to the target raised to the fourth power, r^{-4} ($= r^{-2} \times r^{-2}$). So, a target twice as distant as another target, but otherwise identical, will yield an echo that is sixteen times as weak.

The echo power also depends on the size of the target, in particular, on its area. Hence a spherical target, with twice the radius of another target, but otherwise identical, will yield a radar echo four times as strong since the echo strength depends on the area, hence on the square of the radius, of the target. In mathematical language, the dependence of the echo power, P_{echo}, on these two factors can be expressed as:

$$P_{echo} \; \alpha \; Area_{target}/r^4, \qquad\qquad (5.1)$$

where α denotes "proportional to." If two (nearly) spherical objects, such as the moon and the sun, have the same angular size in the sky, then we can conclude that their respective surface areas divided by the square of their respective distances from us, $Area/r^2$, will be the same. Hence, in relation 5.1, the ratio of the P_{echo}s for the two bodies would depend on the inverse ratio of the squares of their distances from us:

$$P_{echo1}/P_{echo2} \; \alpha \; (r_2/r_1)^2. \qquad\qquad (5.2)$$

Putting the relevant numbers together, we find that were the sun to reflect radio signals like the moon, the strength of the echo received from the sun would be about 150,000 times weaker than the echo received from the moon. In other words, a radar that just barely detected moon echoes would have to be improved in sensitivity by a factor of about 150,000 to (barely) detect sun echoes.

Now we come to the second reason that it's difficult to use radar to measure the distance to the sun: the sun does not have a solid surface and would absorb those signals or reflect them from a nonsolid surface, depending on the wavelengths (or frequencies) of the radio signals transmitted to it. Would reflection from such a vague surface limit the accuracy that could be achieved in determining the astronomical unit in terrestrial units of distance? Yes. But we don't need to direct radar signals toward the sun to obtain a value for the astronomical unit in terrestrial units. Why not? Again, we call on Kepler's third law. Because we measure the orbital periods of the planets with such exquisite accuracy, we know the semimajor axis of each of the planets in astronomical units with about the same accuracy. Thus, knowing the orbits of each planet accurately in astronomical units, we can easily calculate the distance between the two planets accurately in astronomical units at any time. Then, all we need is one measurement between the earth and another planet in terms of a terrestrial unit of distance, and we can determine that distance in terms of astronomical units via a simple calculation. (Actually, the determination is not quite so simple. We do need to determine first, for example, the orbits of the earth and the target planet, plus the planet's radius and shape, to the relevant accuracies.) We therefore do not need to directly measure the distance between the earth and the sun.

Venus, when closest to the earth in its orbit, is the next easiest generally available radar solar-system target to detect after the moon; the required sensitivity is then only a bit over one million times greater than is needed to detect the moon. So how long did that improvement take to achieve? About fifteen years. The remarkable improvement in all aspects of radar systems from antenna size, transmitter power, and receiver sensitivity gave radar the greatest sustained increase in instrumental capability of any of which I'm aware—more than a factor of two per year overall increase in sensitivity on average for nearly fifty years! Our group, then at MIT's Lincoln Laboratory and led by Gordon Pettengill, succeeded in reliable detection of the time delay of radar echoes from Venus in early April 1961, near the time that the earth and Venus made a close approach to each other. Daily observations, made over less than a week, yielded a vast improvement in the determination of the astronomical unit in terrestrial units—by a factor of more than a thousand. What a triumph! It was then realized that the previous uncertainty in our knowledge of the value of the astronomical unit in earth units of length was in reality ten times larger than it was thought to have been.

The radar system used, which had a 25-meter-diameter antenna, was also the first radar to track an artificial earth satellite, Sputnik 1, soon after its launch by the Soviets on October 4, 1957; it is still in use.

Nearly as accurate results for the astronomical unit were obtained from observations made a few days earlier under the leadership of Richard Goldstein, using the Jet Propulsion Laboratory's Goldstone radar and employing a somewhat less accurate technique: measurement of the Doppler effect on the radar signals, an effect discussed later in this chapter. These determinations allowed dead reckoning of Mariner 2 toward its encounter with Venus.

We then knew all distances in the solar system, not only in astronomical units, but with more than sufficient accuracy for navigating in kilometers, for example.

What about the scale of the solar system in terms easily comprehended by a human? An analogy probably does the trick in providing an answer as well as anything. In that spirit, consider the earth shrunk to the size of a piece of pepper, such as is found in grinders on many kitchen tables. If the sun were shrunk by the same factor, it would be about the size of a basketball, and the distance between it and the earth would be about the distance of one end of a football field to the other. Take a moment to imagine this analogy; I think it gives a reasonable feeling for the enormity of the solar system compared to the size of the earth and how empty of objects the solar system really is; the space between objects is extremely large compared to the sizes of the objects in the solar system. Compared to us, of course, the earth's size is huge: the diameter of the earth is about seven million times larger than my height. By comparison, the distance of the sun from the earth is only about twelve thousand times as large as the diameter of the earth.

Distances to Stars

Now we know solar system distances remarkably well in terrestrial units. What about beyond the solar system? The second rung of the cosmic distance ladder enables us to obtain distances to nearby stars. Here we can triangulate, using the earth's *orbit* as a baseline. How exactly do we pull off this trick? We use the concept of parallax: the apparent change in direction when we observe, from two different directions, a relatively nearby object silhouetted against much more distant objects.

Using elementary trigonometry, we can calculate the distance to the apex of a triangle, if we know its base, from measurement of the angle from each end of the base to the apex. We can do the same with stars, where we use the earth's *orbital* diameter as the base. In this case, we measure the angle from each end of the base to the star at the apex of the triangle. We make these angle measurements with respect to background stars, which are much, much farther away than the target star whose distance we wish to determine by these so-called parallax measurements. There is, however, one intrinsic problem: How do we know which stars are close and which are far away? We might think that a brighter-appearing star is closer and a dimmer star farther away. Unfortunately, this rule of thumb is not reliable because stars are not all of, or even close to, the same intrinsic brightness (luminosity). In practice we may use many stars as reference stars and try to determine whether a target star seems to move with respect to all of them or only with respect to some of them. If the picture is mixed, we could try to find other stars nearby on the sky to the target star to test as reference stars, until we find some that don't appear to move with respect to each other; then we can with reasonable confidence assume that they are far enough away to have no observable parallax of their own.

This method of using parallax to determine the distance to stars was tried seriously first in England by James Bradley in 1729. With the highest accuracy then achievable in angle measurements, about 10 seconds of arc, he detected no apparent change throughout the year in the position of a star—that is, no stellar parallax. Bradley did, however, discover aberration, the fact of nature that the direction from which light appears to arrive, for example, from a star depends on the velocity of the earth. This attribute of nature is similar to one's experience walking in the rain: the direction from which, in no-wind conditions, the rain seems to be coming at you depends on how fast you are walking and on the direction that you are walking. Can you think what this result of Bradley's implied about a geocentric versus heliocentric model of the solar system and why?

This proposed parallax measurement of the distance to stars was also realized long ago to be a means to establish definitively whether the earth was moving around the sun or vice versa. If the earth remained stationary while the whole universe rigidly circled around it, then no changes would be observed during a year in the apparent direction of any star with respect to any others. But if the earth were moving relative to the stars and if the stars were at different distances from us, then we might observe a yearly change in the position of the relatively nearby stars with respect to those of the very much more distant ones. If positive results were obtained for the parallax of stars via this technique, and were they repeatable year after year, that would be considered rather incontrovertible proof to (almost!) everyone that the earth orbited the sun and not that the earth was at rest with respect to the remainder of the universe. Bradley's discovery of aberration, of course, pretty well proved to almost everyone's satisfaction that the earth moved with respect to the stars.

If no parallax were detected for any star, what could we conclude? There are two obvious interpretations: either the earth really was motionless with respect to the stars or the stars were so distant that measurement accuracy was insufficient to allow us to detect the motion that parallax measurements would be expected to disclose. It was not until 1836, more than a century after Bradley's attempt, that parallax-measurement instrumentation and methodology had improved sufficiently to yield a positive result. Friedrich Wilhelm Bessel, a famous mathematician and astronomer, succeeded in determining the parallax of a bright star, 61 Cygni, of about 0.3 arc seconds.

How was this parallax measurement converted to one of distance? For this purpose, a new unit of distance was introduced, the parsec. This unit, a combination of "*par*allax" and "arc *second*" (coined in 1913 by Herbert Hall Turner), is the distance at which the radius of the earth's orbit, more accurately the astronomical unit (au), subtends an angle of one arc second (fig. 5.1). Note the possible confusion of a factor of two: the parsec is defined using the *radius* of the earth's orbit around the sun, but the measurements, to give the highest accuracy, make use of the *diameter* of the earth's orbit, which is of course twice as large as the radius.

One now widely used unit of distance in astronomy is the parsec. It is not the only such unit. The parsec reigns alongside the light-year (1 parsec = 3.26 light-years). The light-year, as you might guess, is the distance light travels in vacuum in one year. Both are often thought of a bit loosely: for example, what "year"? (There are various definitions, such as from vernal equinox to vernal equinox; only one should be used for the definition of a light-year.) Measurement of distance outside the solar system is not nearly accurate enough for this looseness in usage to yet affect any results importantly.

From the surface of the earth, the "twinkles" introduced by the earth's atmosphere limit the distance to which the parallax method can yield reliable results to the order of about 100 parsecs. However, with a spacecraft, dubbed Gaia, launched in December 2013 by the European Space Agency,

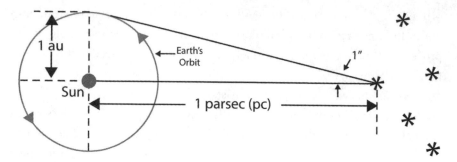

Figure 5.1. A parsec (a unit of distance) depicted in terms of parallax. Note: this figure is *not* to scale; for one example, the distant reference stars are not shown very far away from the star at the apex of the triangle used to define the parsec, as they should be to form a realistic background of objects that do not discernibly exhibit any parallax. These distant stars are used to measure the angle between the apex star and them from each of the two ends of the base of the triangle, formed by the radius of the earth's orbit. Courtesy of David Shapiro.

the parallax method is being used to extend measurements of stellar distances to the order of 10,000 parsecs, with accuracy degrading only slowly with distance from best accuracies of order 10 microarcseconds, thus revolutionizing such measurements. One point of information: the method being used by Gaia to determine the distances to stars came under suspicion. Results from a prior European spacecraft, Hipparcos, using the same approach, were shown to be inconsistent with results from another, more reliable, technique, very long baseline interferometry (VLBI, certainly a mouthful, to be defined and discussed in chapter 10). Investigation showed conclusively that the VLBI result for the distance to the Pleiades (a star group) was accurate and that the Hipparcos result was in error. The Gaia result for the Pleiades agreed with the VLBI result and not with the Hipparcos result. No reason has yet been found to explain the discordant result from Hipparcos. This story illustrates that science is not always straightforward but often takes unexpected routes. However, the truth always triumphs in the end, or so the dogma states.

Distances via the Period-Luminosity Relation

How do we determine distances to objects farther out in the universe than can be reached via parallax measurements? A clever method arose via a combination of brains and luck. A bit over a century ago, the Harvard College Observatory (HCO) had a telescope in Peru that was used to observe stars visible only from the southern hemisphere. One region of that southern sky, called the Small Magellanic Cloud (SMC), had a large concentration of stars, presumably gravitationally bound to each other, with all at roughly the same distance from us. The observers, located in Peru, made repeated observations of the stars in the SMC. The large glass photographic plates that contained these images of the southern sky were sent back to HCO for detailed analysis. At that time, HCO had a set of sharp computers to carry out these analyses—not Apple or Samsung or Dell or even IBM computers, which didn't then exist. These were all women who computed, thus "computers" (fig. 5.2). In that era, opportunities for women in science were scarce to nonexistent. Edward C. Pickering, then the HCO director, hired many of the best and the brightest, and they did superb research. (Dava Sobel published a book on these "computers," entitled *The Glass Universe*,

Figure 5.2. Computers at work at the Harvard College Observatory, around 1900.

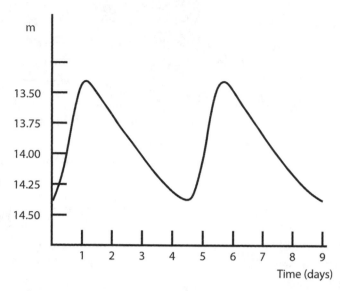

Figure 5.3. A short segment in time of nine days showing the brightness variation in units of magnitudes of a type of variable star discovered by Henrietta Leavitt. Magnitudes ("m" on y-axis) are a logarithmic representation of the brightness of stars; the higher the magnitude, though, the lower the brightness, so that in this scale a brightness of 13.50 magnitudes is about 2.5 times brighter than a brightness of 14.50 magnitudes. Adapted by David Shapiro from ESA/Hubble & ESO Astronomy Exercise Series, CC BY 4.0.

in 2016; it became a best-seller.) One of these computers, Henrietta Leavitt, noticed in examining her data on variable stars that there were some whose brightness varied with time in a periodic manner (fig. 5.3). Moreover, she noticed a totally unexpected relation: the period of the brightness variations of these stars, called Cepheid variables, was related to their mean brightness, or equivalently their maximum or minimum brightness. In particular, the greater the average, or mean, brightness of the Cepheid, the longer was the period of its brightness variation.

How could this be? Clearly these Cepheids could not be at random distances from the solar system; they all had to be at about the same distance for observation of such a relation to even be conceivable. Why? The brightness that we detect—the apparent brightness—for a star will vary as the inverse square of the distance of the star from us. (The distinction between a star's apparent brightness—the brightness of a star as it appears to us on the earth—and a star's luminosity—the intrinsic brightness of the star, a measure of how much energy it radiates, or emits, per unit time—is important to keep in mind.) Hence, such a period-luminosity relation (PLR) would be exceedingly unlikely to be discovered were the Cepheids at random distances from us, as opposed to being within a relatively narrow range of distances from us.

Since the period of variation of a Cepheid's brightness is observable independent of its distance from us (so long as it is bright enough for us to be able to detect its brightness variations), then we can use these variations, or rather the period of these variations, to infer its distance (see fig. 5.3). We do so by comparing the observed mean, or average over a cycle, of its brightness variations, or of the maximum or minimum of the apparent brightness of its variations, with the corresponding quantity of a Cepheid variable, which has the same period of brightness variations and is at a known distance from us. Thus, we set the ratio, say, of the two mean brightnesses equal to the ratio of the inverse squares of the distances for two Cepheids of the same period of brightness variations:

$$B_1/B_0 = r_0^2/r_1^2, \tag{5.3}$$

and, via a little algebra, we can solve for the distance, r_1, to the Cepheid whose distance we seek:

$$r_1 = r_0 (B_0/B_1)^{1/2}, \tag{5.4}$$

where B_0 and B_1 represent, respectively, the mean brightness of the reference Cepheid and the one whose distance we seek, and r_0 and r_1 are their corresponding distances from us. We can quickly check that this ratio of the Bs is right side up and not upside down, by noting that the dimmer the unknown Cepheid is, the farther away it must be; hence B_1 is in the denominator, making the fraction larger the smaller the denominator. Note that in the above two formulas, the brightnesses are expressed *not* in magnitudes but in units such that the brighter the object, the higher the value of B, in direct proportion.

The basis for equation 5.3 may seem obscure. Let me try to clarify: since the two Cepheids have the same period of brightness variation, they have the same intrinsic brightness if they obey the PLR. They differ in their apparent brightness, however, because they are in general at different distances from us. But their apparent brightness varies with the inverse square of their distance from us. By multiplying the brightness of each Cepheid by the square of its distance from us, we compensate for the effects of the inverse square of the distance. The result of that multiplication is thus independent of distance and, hence, is the same for the two Cepheids, which have the same intrinsic brightness, or luminosity. Because the periods of variation of their brightness is the same and they presumably obey the PLR, we have thus justified equation 5.3. I hope that this explanation is clear, if a bit redundant, although it may take two readings to appear evident.

So here we have an *indirect* method of determining *relative* distances. To determine *absolute* distances, distances say in parsecs, or light-years, we have to calibrate the relation—that is, we have to find out what distance goes with a specific, say, average brightness of some Cepheid. How can we carry out this calibration? One obvious means is to find a Cepheid variable, whose average brightness we've measured, close enough to us that we can measure its parallax and hence its distance. That calibration was in fact done. But strange as it may seem, uncertainty in this calibration is the limiting factor in the determination of distances via this method, even today, a bit more than a century after Henrietta Leavitt discovered the PLR. Of course, the calibration is more accurate now than it was a century ago, yet it remains the limiting factor in the accuracy of this indirect distance-calibration technique. The Gaia spacecraft should improve the pre-Gaia calibration dramatically, by more than a factor of ten, over the next years, through increasing accuracy of parallax determinations and extensions of them to distances up to one hundred times greater.

Note that for all distances greater than those for which we can use parallax measurements, we need and use *proxies*, starting with the period-luminosity relation. The proxy method based on this relation constitutes the third rung of the cosmic distance ladder. Too long overdue, though, is explicit recognition of the discoverer in the name of this relation. It should in my opinion be officially named Leavitt's period-luminosity relation or an equivalent.

The Redshift-Distance Relation

What is the fourth rung of the cosmic distance ladder, and why do we need it? Let us address the second question first: Cepheid variables are not luminous enough to be observed in very distant galaxies. To set the stage for this next rung, we will, as is our stated custom, take the historical route.

We start with a question. What did we know a bit over a century ago about the size of the universe? Not too much. At the time, astronomers argued over whether the Milky Way constituted the whole of our universe or if it was just one of many galaxies. Our telescopes were then good (big) enough to disclose hazy patches of light—nebulae—in various directions from us. These patches seemed to be distributed more or less randomly on the sky. By contrast, stars seemed concentrated in a band, the Milky Way, which indicated that the shape of our galaxy resembled that of a pancake. If these hazy patches of light, also dubbed island universes, were part of our galaxy, how could they be distributed approximately uniformly on the sky instead of like the pancake? Moreover, where were we, and our sun, situated with respect to the center of our galaxy? And how could we find out?

Because the nature of the debate was so fundamental, with our knowledge of the makeup of the structure of the entire universe hanging in the balance, the prominent astronomer George Ellery Hale suggested that a debate be held at the U.S. National Academy of Sciences between two astronomers, each an articulate spokesman for *his* point of view (no women were allowed at that time to reach the necessary stature). The debaters chosen were Harlow Shapley, then on tap to succeed Edward C. Pickering as director of the Harvard College Observatory, and Heber Curtis, a much older astronomer and more accomplished orator from the famous Lick Observatory in the San Francisco Bay Area. The debate took place on April 26, 1920; no transcript or recording was made, but a most probably, nay doubtless significantly changed, version of the debate was printed about a year later. This printed version gave the position of each protagonist on each of the two main questions. On the main issue—one galaxy or many—Curtis had by far the better argument, for example that randomly distributed island universes could not be understood as residing within an overall pancake-shaped universe; Shapley's defense of there being only a single universe was based in part on a claim (later shown to be false) that one island universe was observed to move in angle on the sky by so much that, had it been outside the least stringent of the sizes then preferred for the Milky Way's size, it would be moving at a speed greater than c and, hence, in conflict with special relativity. On the other hand, Shapley failed to explain the observed distribution on the sky of the island universes.

As for our distance from the center of the galaxy, Shapley's argument was the more persuasive placing us at about 10,000 parsecs from the center, several times closer than did Curtis. No matter who, if either, was correct here, note how our believed special place in the universe continued to degrade: we were now found to be even far removed from the center of our own galaxy, whether unique or one of many.

While arguments over answers to these basic questions took place, the underlying assumption was that the universe, overall, was static. Einstein, whose theory of general relativity, he thought, should provide a model for the behavior of the universe, was puzzled. How could the universe remain static? It should collapse in on itself due to all matter attracting all other matter (as, indeed, in Newton's theory, too), thus making a static universe unstable against collapse: such a universe over time would collapse. But if the universe were, instead, really static as most, or at least many (I took no poll), astronomers of that time claimed, Einstein's theory of general relativity could not possibly be a good model for the whole universe. Nevertheless, this theory seemed to work quite well for shorter spatial distances, as in our solar system.

What to do? Einstein had been in tight theoretical jams before. His solution here was brilliant. He proposed adding to his equations an extra term on the geometry side, which term had no discernible effect over small spatial scales but acted like a repulsive force over large spatial scales. Operating like an antigravity force of just the correct magnitude, the new term was determined by a constant whose value depended on the particular characteristics of our universe. Adding this term to the geometry side was a drastic departure from his original theory, which boasted of having no arbitrary constants. Einstein termed this proposed new constant, the coefficient of this new term, the cosmological constant.

Meanwhile, the theoretical application of Einstein's theory of general relativity to cosmology—the structure and evolution of the universe—was bubbling with activity. Alexander Friedmann, a brilliant young scientist in the Soviet Union, had in the early 1920s deduced that an *expanding* universe was a possible solution of Einstein's equations of general relativity. Similarly, Georges Lemaître, a Belgian cleric, independently reached similar conclusions in 1927. Shortly thereafter, Edwin Hubble, now of Hubble space-telescope fame, set out to use the most sensitive telescope then available, the 100-inch-diameter Hooker reflecting telescope (the mirror is 2.54 meters, or 100 inches, in diameter), on Mount Wilson just north of Pasadena, California, to measure the distances to some of the island universes. He succeeded in the late 1920s (see below), using Cepheid variables, to measure distances to some island universes, thereby showing that they were indeed separate galaxies, up to several *million* light-years from us.

For Hubble's analysis of the light from the galaxies that he observed, he also used their spectra, which had been obtained mostly if not wholly by the American astronomer Vesto Slipher a decade or so earlier. What were these spectra? Let us digress to provide a brief background in optical spectra because they play a key role in much of what follows in this first section of the book. We start as usual with some history: the spectrum of the sun was observed in 1813 by German physicist Joseph von Fraunhofer (confirming and extending 1802 observations by W. H. Wollaston). Fraunhofer observed that when sunlight passes through a clear prism, the resultant colors shining on a screen—the spectrum (intensity versus frequency or color)—contained dark, thin lines at certain places. Fraunhofer was unsure what these lines meant. Later scientists discovered that when an electrical current was passed through a gas of a particular atomic substance (an element), such as pure oxygen or pure hydrogen, and the light consequently emitted by the substance was then passed through a prism, or equivalent device, the light showed lines at very specific frequencies, or colors, unique for each particular substance. The pattern seen was characteristic of the element: in the laboratory, the lines for any particular element always appear in the same place—that is, they always appear at the same frequencies or, equivalently, colors. Each element has a distinctive spectrum, which thus acts as a fingerprint for the element. These so-called bright emission lines each corresponded to a dark (=absorption) line in the solar spectrum, where the elements in the outer part of the sun were absorbing the light coming from the hotter, more inner part of the sun—puzzling at the time.

What was revealed by Slipher's spectra of the galaxies that Hubble observed? These spectra showed characteristic lines, identifiable with specific elements, but with one major difference: the positions of these lines as seen in elements in the sun were shifted in frequency compared to the lines of the spectra of the corresponding elements when observed in the laboratory. Why were the lines shifted? In which direction were they shifted? And by how much were they shifted? Here we require yet another digression to provide the background needed to explain the answers: a short discussion on the speed of a wave and the Doppler effect.

The Speed of a Wave

Back in 1844 in Vienna, Christian Doppler, while studying a double star with his telescope, observed that the pattern of the light shifted back and forth in frequency, which he deduced was due to the orbital motion of each star about the other, this last being the definition of a double star. This characteristic of light is very similar to the effect we are all familiar with for sound. For example, as a train approaches us, its whistle appears to have a higher pitch than after it passes us and is moving away. This change in pitch, or frequency, of the sound is called the Doppler shift, after you can guess whom. In fact, one could argue that it should be called the Rømer effect, since Ole Rømer first uncovered this shift nearly two hundred years earlier and applied it in his demonstration that light traveled at a finite speed. How can we express this shift mathematically for the light analogy, as a function of the velocity of the source with respect to us? First, we need a model for the light. One that works well for this application is a wave model (fig. 5.4). Think about a water wave. The peak height of the water corresponds to the crest. The distance between one crest and the next is the wavelength, λ, and half the vertical distance from the bottom (trough) to the top (crest) is the amplitude, a, of the wave. If the wave travels horizontally, at any

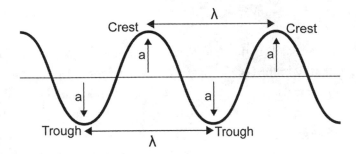

Figure 5.4. Wave motion (see text). Courtesy of David Shapiro.

point in space that the wave passes, the material there will go up and down periodically.

How many times the water goes through a wavelength per unit of time is called its frequency. If we think now about a wave of light, we can find its speed, c, which is thus equal to the product of its wavelength and its frequency:

$$c = \lambda \times f. \tag{5.5}$$

We can perhaps make this relation clearer by thinking of a simple case and then generalizing. Suppose we are standing in a place watching a wave go by, and in one second one whole wave goes by, say crest to following crest. The speed of that wave would be the wavelength multiplied by one, the number of wavelengths that went by in one second—namely, one in this case. If two whole waves went by in one second, then the speed of travel of the wave would be twice what it was in the first example, given that the wavelength of the wave were the same in the two examples.

What about the Doppler shift? Its derivation is a bit involved; therefore, I present here just the result. Thus, if the source of the signal is moving toward the observer, the frequency received by the observer will be given by:

$$f \approx f_0 \left[1 + (v/c) \right], \tag{5.6}$$

where v denotes the speed at which the source is approaching the receiver of the signal, where v is much less than c, the speed of light, and the $\approx$ means approximately equal to. If the source is moving away from the observer at the same speed, the only change to equation 5.6 is to change the plus sign to a minus sign.

But wait a second. Didn't Einstein postulate, with observations having upheld it, that light travels with speed c independent of the velocity of the source with respect to the receiver? Indeed. The implication here is that the wavelength, denoted by the Greek letter λ, must decrease if the frequency, f, increases, so as to preserve the speed: $f \times \lambda = c$, just as $f_0 \times \lambda_0 = c$.

We assume that the universe everywhere obeys the same rules that we find obeyed here on earth. This is a fundamental assumption. Thus, if we observe the lines of the spectra of distant objects to be the same as we observe in the laboratory, save for all being shifted proportionally by the same amount, we draw the conclusion that this shift is due to the difference in the line-of-sight velocity of the emitting object(s) and us. There exist other possible causes, but none has so far stood up to probing analysis.

Returning to Edwin Hubble: he was able to deduce distances out to several million light-years, based in part on observations of Cepheid variables in these distant galaxies. He could see them well enough because he had use of a more powerful telescope than had been previously available. These results showed rather conclusively that these island universes lay far beyond the then-believed confines of the Milky Way, turning the tide of astronomical opinion firmly in the direction of island universes being other galaxies, lying outside the confines of the Milky Way. Thus was our self-importance—our privileged place in the cosmos—knocked down yet another notch, to a position far from the center of a run-of-the-mill galaxy. Finding firm evidence for intelligent life (see chapter 22) elsewhere in the universe might be the next and final knockdown for our prior thoughts of holding a privileged place in the universe.

Hubble, using Slipher's spectra from the light from these galaxies whose distances Hubble had estimated, apparently noticed a remarkable relation: the distances were proportional, roughly, to the observed redshifts of the spectra, in other words indicating that the galaxies were moving away from us at velocities proportional to these distances (fig. 5.5). That is, each spectrum was shifted to the red; the elements, identified in the spectra by their characteristic lines, had these lines all shifted proportionally to the redder

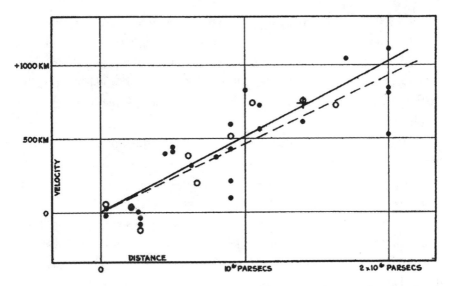

Figure 5.5. Edwin Hubble's classic result of 1929 for velocity versus distance of nearby galaxies in units of megaparsec (10^6 parsec). Ignore the differences between the hollow and filled
circles and the solid and dashed lines. (Note that the graph omits "per second" from the label
"KM" on the vertical axis, which shows the speeds, in kilometers per second. In other words,
there is this mistake in the labeling of Hubble's graph.) Edwin Hubble, "A Relation between
Distance and Radial Velocity among Extra-Galactic Nebulae," *Proceedings of the National
Academy of Sciences* 15, no. 3 (1929): 168–173. Courtesy Carnegie Institution for Science.

(longer wavelength) part of the spectrum. Their frequencies were lower
than those measured in the laboratory, indicating that these sources were
moving away from us, at speeds directly proportional to their distances from
us. What interpretation follows from this remarkable observation? Well,
suppose we were to uniformly stretch an elastic band or uniformly blow up
a balloon. How would evenly spaced points on the band or evenly spaced
dots on the balloon move with respect to each other as the band's length or
the balloon's radius increased uniformly? The distances between neighboring, next neighboring, and so on, points would increase at rates proportional
to this separation.

Thus, Hubble's observations, whether wholly independent of, or at least
partially inspired by, Friedmann's and/or Lemaître's work, demonstrated
that the universe was expanding. This seemed an inescapable conclusion to

be drawn from his redshift-distance relation, were it to be true in general. It is interesting that, in his findings, published in 1929, Hubble did not draw this conclusion. He was cautious, but why he decided to draw other (albeit tentative) conclusions and write nothing at all about this one is not possible for me to say reliably. Could it have been that he did not recognize this possibility at that time?

One could more easily deduce the redshift from the light from the atoms of a specific, common element distributed all over a galaxy than from the light from a single separate Cepheid variable—more total light. Thus, one could measure redshifts of whole galaxies at distances much greater than one could measure the brightness of a single Cepheid variable, even if one could distinguish the light from individual Cepheids. This redshift-distance relation (RDR) thus took up the next, the fourth, rung of the cosmic distance ladder in the following form:

$$v = Hd, \tag{5.7}$$

where v represents the velocity, or speed, of a galaxy's motion away from us and d its distance from us. Here H is a constant of proportionality. Whether this constant remains fixed as d increases significantly is a topic we will return to in chapter 7. This constant is named after guess who, despite, as we will describe in the next chapter, Lemaître having published a value for it two years earlier. Thus, many astronomers think that this constant should be written as "L," not "H," and named Lemaître's constant, rather than Hubble's constant. But, it seems, Hubble was an excellent publicist, a characteristic to my knowledge not exhibited by Lemaître.

A difficulty with this rung of the cosmic distance ladder is the spread in velocities due to the motions of the stars within galaxies, which to some extent smear out the overall motion of the galaxy. In addition, there are the so-called peculiar velocities of galaxies, which are overall galaxy velocities that differ from the general, uniform velocity of expansion. These peculiar velocities are similar to variations in speeds with which different people, analogous to different galaxies, leave a burning building at a specific instant, although, of course, this analogy with the small variations in the RDR breaks down right away with the main part of this relation.

Having probably exhausted ourselves on this fourth rung, we move on to the fifth and last one.

The Fifth Rung of the Cosmic Distance Ladder

The fifth and last rung of the cosmic distance ladder that we consider is the rung that takes us beyond the redshift-distance relation, with possibly more accuracy, due for example to the absence of the smearing effect of the relative motions of individual stars in the composite result from observing the light from an entire galaxy. This fifth rung is based on *supernovas* as distance indicators. What is a supernova? A supernova appears in the sky all of a sudden as an intensely bright object, pointlike, but with a brightness that can—for a short time—equal or exceed that of its entire galaxy. Its brightness increases greatly during its first few days and then decreases more slowly, say over a month or two, until one can no longer distinguish it, if it occurs in a distant galaxy. There are ways of characterizing a number of different classes of supernovas that can be distinguished from each other by observations, based on the spectra and the rises and falls in brightness with time of the supernovas. There exists at present, however, no completely satisfactory model for supernovas of any class.

Scientists consider many supernovas (SNs) to be the final stages of a star that has burned all of its nuclear fuel and then gravitationally collapses under its own mass, setting off a huge explosion, the heart of a supernova. By studying supernovas carefully, one type, called SN 1a, has been calibrated pretty well as a standard candle.

What is a standard candle? It is a descriptor for a type of object whose luminosity is known and is very accurately the same for all members of that type. Hence, one can count on this constancy and use the apparent brightness of any member of the class to determine its distance from us via the inverse square law, save for possible effects of any intervening material (see chapter 7 for a short discussion of this possibility). Of course, to be used to determine distances, the distance of at least one such supernova must first be determined by some other means. This calibration for supernova distances relies on at least one member of the class being identified in a nearby galaxy where a lower rung in the cosmic distance ladder can be used to reasonably reliably determine the distance to the supernova.

This supernova standard candle thus provides a believed-to-be reliable distance indicator out to distances of several billion parsecs. We leave further discussion of it to chapter 7.

An Overall Ladder Property

The cosmic distance ladder has another property unexpected by anyone familiar only with ordinary ladders. This property is unobservable on an ordinary ladder: the farther up the cosmic distance ladder we go—that is, the greater the distance we look out—the farther back we see in time. The light that we now see left the object we are observing at an earlier time, earlier by the length of time taken for the light to get from that object to us. Thus, we will see an object 1,000 light-years away as it appeared there a thousand years ago. Succinctly: *Looking out in space means peering back in time.*

Having completed our distance tour and completed the construction of our cosmic distance ladder, we turn to its use in the following chapters to probe more deeply several aspects of modern cosmology.

SIX
===

The Cosmic Microwave Background

In 1927, Father Georges Lemaître, gifted scientist and Belgian priest, was apparently the first to determine the redshift-distance relation and to note its implications of an expanding universe. Edwin Hubble, due in part to his penchant for self-promotion, is most often credited with this discovery, but the publication history shows that Lemaître was first. He was also first to estimate the rate of expansion, now universally called the Hubble constant, not the Lemaître constant. This is the way history sometimes goes . . . . But there has now been partial recognition of Lemaître: the International Astronomical Union has officially changed the name of the redshift-distance relation to the Hubble-Lemaître law. As a reminder, this relation states that distant galaxies recede from the Milky Way at a speed proportional to that distance. You can rather easily figure out that this relation implies that the universe is expanding. An expanding universe has a profound consequence: if the universe has continuously expanded, we can, as a *gedanken* (thought) experiment, realize that the universe must have been smaller in the past. One can in principle trace back to the time when the universe was very small, as Lemaître did, although he preferred to infer that the universe was, at the start of its expansion, not a single point, indescribably smaller than an atom, but a very small size nevertheless—just a giant atom. Currently, the prevailing theory is that the universe started its expansion from a point size. This is the so-called Big Bang theory. The scientist Fred Hoyle, who was a main proponent in the mid-twentieth century of a rival theory, the Steady

State theory, now discredited, apparently invented the phrase "Big Bang" in the late 1940s as a derisory label. It stuck, as a favored label.

Given the fundamental nature of this Big Bang concept, what kind of other observational support might we seek to test it? Answering this question is the goal of the remainder of this chapter. First, let us note that by tracing back to the Big Bang, we are estimating the age of the universe. Pretty heady stuff! However, not until nearly two decades after Lemaître's work did anyone seriously try to calculate the observational consequences of a Big Bang. Without delving deeply into the presently accepted mathematical model of an expanding universe, I will simply sketch that model qualitatively and cursorily: light, or radiation, coming out from the original Big Bang explosion was extremely hot and intense. Yes, this radiation had a temperature, as we discuss below (see, too, *The First Three Minutes*, by Steven Weinberg, for a far more expanded treatment). By the way, radiation encompasses not only visible light, that which we can discern with our eyes, but all wavelengths (and frequencies) of the spectrum, from radio waves through X-rays and beyond. This radiation from the Big Bang explosion was predicted to have been equally intense in all directions (see below for small exceptions), since it presumably came from a (very nearly) spherically symmetric, pointlike explosion. This situation quickly evolved into one that contained, in addition to the light, elementary particles (see chapter 11) created from some of the light. This then opaque soup continued expanding and cooling until about 380,000 years later, according to current models, after which atoms formed and the light continued on its merry way unimpeded to expand and to cool. (Why cooling? A simple, short, explanation: the energy density of material, directly related to its temperature, decreases as the material expands into a larger space.) What would be the temperature of this radiation here now? Not until the late 1940s were calculations published for the expected current temperature from this cooling radiation that had, according to current theory, originated in the very early universe. Ralph Alpher, Robert Herman, and George Gamow, about whom, again, more later, estimated that the current temperature of the cooling radiation from the aftermath of the Big Bang would now be about 5 Kelvin. That is, 5 degrees above absolute zero, the lowest temperature on the Kelvin scale and considered to be the lowest possible temperature, consistent with our ideas as to what temperature means. Degree spacings in the Kelvin scale are the same as those on the Centigrade scale. Only the origin—the zero point—on the scale is different; this point is set at the freezing temperature of water on the Centigrade scale and at so-called absolute zero on the Kelvin scale.

Did this prediction prompt experimentalists to seek evidence of this radiation? Apparently few, if any, tried, and none succeeded. Why? At a temperature of 5 Kelvin, the peak of the radiation—the wavelength of the radiation that had the greatest-intensity emission (see the discussion of black-body radiation below)—would be at a wavelength of 0.6 millimeter and a corresponding frequency of 500 gigaHertz, in the radio part of the spectrum, and with so small an amplitude that it would have been very difficult to pick out with equipment then available. Nonetheless, I suspect that the true explanation more likely involves a combination of the difficulty of the measurement and the fact that cosmology was not then very fashionable with astronomers, in general, or with radio astronomers, in particular. Radio astronomers, mostly with backgrounds in electrical engineering and/or radio physics, were not very numerous and were the ones who would perforce have made such microwave measurements. Physicists only began to seriously infiltrate astronomy in the late 1950s and early 1960s. What prompted this infiltration was most likely the startup of the Space Age, initiated by the launch by the Soviet Union in 1957 of Sputnik 1, and the corresponding opening up of intriguing new problems to pursue, along with the many new positions and funding opportunities that opened in which to pursue these problems, as the United States tried to catch up to the Soviet Union. The fact remains, though, that the 5 Kelvin prediction seemed to have been ignored. In some ways this was a situation parallel to the (near) ignoring by British and French astronomers of the predictions of Adams and Le Verrier, respectively, of the existence of the planet Neptune.

Black-Body Radiation

Physicists have known for over a century that the light, or radiation, coming out of a furnace has a spectrum characteristic of those emitted by so-called black bodies—that is, bodies that absorb all of the light that strikes them, reflecting none at any wavelength. (A black body is, though, not the same as a black *hole*.) It may seem somewhat paradoxical, however, that a black body, which absorbs all external light that impinges on it, nonetheless *emits* light of its own. In particular, nature behaves, as our observations disclosed, such that these black bodies emit light whose characteristic spectra depend on only one parameter, the temperature of the body. The spectra are independent of the body's composition, density, shape, whatever; these spectra depend only on the body's temperature. This fact was known well before a satisfactory model of the spectrum was developed.

The spectrum of every black body also possesses another characteristic: the intensity of the emission is zero at zero wavelength, rises to a maximum monotonically—that is, continually rises to a maximum—and then falls, again monotonically, to zero at infinite wavelengths. The same pattern is exhibited when we describe the spectrum in terms of frequency instead of wavelength: the intensity starts at zero and ends at zero. For a black-body spectrum corresponding to a black body at any temperature, T, the wavelength, λ_{max}, at which the intensity of the radiation reaches its maximum satisfies an equation called Wien's displacement law, after its discovery in 1893 by Wilhelm Wien:

$$\lambda_{max} T = 0.3 \ cm\text{-}K, \qquad (6.1)$$

where "-" indicates a dash, not a minus sign—that is, showing that the units of 0.3 are centimeters Kelvin. For example, at $T = 5$ K, the wavelength of maximum intensity of black-body radiation is predicted to be at 0.06 centimeter, as can be inferred from equation 6.1 and as was noted earlier in this chapter. As illustrated here, the units for the points on the horizontal axis are nanometers (nm), which are 10^{-9} meters (fig. 6.1); thus, 0.06 centimeter = 0.06×10^7 nanometers = 600,000 nanometers, *way* off the right-hand part of the figure.

Many (most?) people find black-body radiation a difficult concept to grasp. Understanding its model is indeed difficult, but just accepting that nature behaves in this manner, as we observe through experiment, shouldn't be so difficult—it is no more bizarre than that, when one lets go of a fork, it

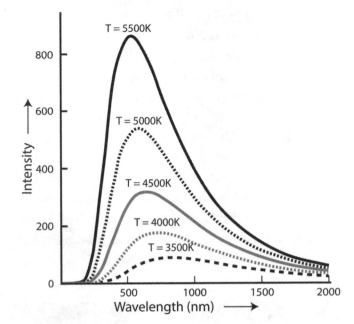

Figure 6.1. Black-body spectra. Note that the intensity curves for the different temperatures do not cross each other and that the peak of the curve moves to lower values and longer wavelengths as the body's temperature decreases. Figure by Unc.hbar.

falls to the ground. It is just that, having been children, we have been expert at dropping things from before we could even talk. We therefore feel more comfortable with this behavior of nature. At least that's my view; I have, though, essentially no supporting evidence.

Note that visible light—wavelengths to which our eyes are sensitive and our atmosphere transparent—is where the sun emits most of its light; its surface temperature is near 6,000 Kelvin. Coincidence?? I suspect not!

The Discovery of the Cosmic Microwave Background

Not until the mid-1960s was a serious attempt initiated to detect this microwave radiation, dubbed the cosmic microwave background (CMB). This attempt was not exactly a slam dunk; despite more than fifteen years of technological advances since the appearance of the paper predicting the CMB, it was still a challenging experiment to detect such weak radiation: all of the other sources of radiation (noise for this experiment) had to be accounted for with the total uncertainty well under the anticipated 5 Kelvin from the CMB. A Princeton University group, under physicist Bob Dicke, nonetheless pursued the construction of the needed equipment.

While the Princeton group was developing its overall instrumentation package with which to try to detect the CMB, in the nearby New Jersey location of Holmdel, two freshly minted young physics Ph.D.s, Arno Penzias and Robert (Bob) Wilson, were working on a very sensitive horn antenna (fig. 6.2). They were working at Bell Laboratories, a part of the American Telephone and Telegraph Company (AT&T), which at that time, as a public utility, had a monopoly on telephone communications in the United States. As an aside, I can say unequivocally that I've never known anyone who was sharper, was more careful, or had higher professional integrity than Bob.

Figure 6.2. The Bell Laboratories' horn antenna that led to discovery of the cosmic microwave background in 1965. Reused with permission of Nokia Corporation and AT&T Archives.

They were trying to determine all sources of sky noise to be able to avoid them so as to make the horn more effective for communications via satellites. (A horn antenna, by the way, provides especially good shielding against unwanted terrestrial signals.) A bit simplistically, we could say: the more reliable the reception, the more profit AT&T could make.

Hence, this project was viewed as an important one for Bell Labs. Making careful measurements all around the sky, day and night, the two physicists located all sources of signals at the microwave frequencies that they were examining, a band near 4 gigaHertz (four billion oscillations per second, corresponding to a wavelength of about 7.5 cm; recall that the speed of light equals its wavelength multiplied by its frequency: $\lambda f = c$). However, they found a residual signal whose origin they could not determine; it had an equivalent temperature of about 3 Kelvin and was uniformly distributed around the sky. What could this radiation be? They examined everything they could think of, including pigeon droppings on the horn, which they meticulously cleaned once discovered. Still this background noise did not disappear. They were mystified. One of them, Penzias, was talking with Bernard Burke at MIT and told him about this mysterious noise, present in every direction the horn was pointed. Burke, being aware of the Princeton University group's budding efforts to detect the CMB, told Penzias about their program. Penzias then contacted the Princeton University group, and the two groups agreed to submit papers for publication, which could be "back to back," with the Bell Labs group publishing its apparent detection of the CMB at a temperature of 3 Kelvin and the Princeton group describing the theory behind this detection and its providing evidence for a "hot Big Bang."

Again we ask the question: Case closed? If not, why not? Clearly it was not closed: measuring the intensity of the radiation at only one frequency doesn't establish that this radiation has the predicted CMB spectrum of a black body (see fig. 6.1). Also, although the Bell Labs scientists were very careful, this measurement was pushing the state of the art and so could have had some subtle errors not easily discerned.

The Bell Labs detection of the CMB, the very first direct observational evidence supporting the Big Bang, ignited the radio astronomical community. Those who had relevant expertise and equipment rushed to make their own measurements. These were, however, all on the low-frequency (long-wavelength) side of the peak of the presumed black-body spectrum; none was on the other side of the peak. However, it was soon realized by two astronomers, George Field and, independently and slightly later, Patrick Thaddeus, that in the late 1930s molecules of cyanide had been observed in relatively nearby interstellar space and appeared to be in an environment that had a temperature of about 2.7 Kelvin. No one knew how that could have come about. What could have supplied the energy in interstellar space to enable cyanide molecules there to reach this 2.7 Kelvin temperature? Now it was clear: the CMB radiation. The case, although not quite closed, was fairly convincing. The CMB existed and provided powerful support for the concept of the Big Bang. As a result, in 1978 Penzias and Wilson were awarded the Nobel Prize in Physics.

Spatial Fluctuations in the Temperature of the CMB

There was still great interest in observing the CMB from space. Why? Had the CMB been perfectly uniform in all directions—that is, had this black body been perfectly isotropic, with no deviations from isotropy (save for quantum effects, which we will not pursue)—then one could conclude that we wouldn't be here to observe it. To form structure in the universe there had to have been some inhomogeneities in the initial explosion, assuming of course that our current theory of the Big Bang is correct—not a sure thing. These inhomogeneities would be small, though. From the earth's surface, it was a great challenge to detect the CMB at all. To detect deviations from isotropy required, at the then level of technical development, going to space to observe: effects on the CMB signals of fluctuations in the density of our atmosphere, for example, make virtually hopeless attempts to observe these tiny effects of Big Bang inhomogeneities from the surface of the earth.

About a quarter century after the Bell Labs detection of the CMB, in 1989, the first follow-up space experiment was launched, called COBE (*cosmic background explorer*). COBE measured the CMB and found that it had the most perfect black-body spectrum ever observed up until that time. Two major discoveries followed.

First, COBE detected the motion of the solar system with respect to the CMB. How did this detection come about? The CMB defines a reference frame: the temperature is the same in all directions, save for the aforementioned (small) irregularities that gave rise to structure in the universe and thence, eventually, to us. But what happens if we observe the CMB from a moving platform—that is, one moving with respect to the CMB? Then, as we discussed in chapter 5 for the Doppler shift, the frequency of the radiation we receive will increase in the direction in which we are moving and decrease in the opposite direction, with intermediate values detected in the other directions. Thus, for example, the peak of the spectrum observed for the CMB would shift toward higher frequencies in the direction the spacecraft was moving with respect to the CMB and to lower frequencies in the opposite direction. The temperature of the equivalent black body would be shifted similarly: the temperature of the CMB black-body radiation would appear to increase the most in one direction and decrease the most in the opposite direction and have intermediate values in the directions in between these extremes. Measurements made from the ground, due to the aforementioned atmospheric effects, were not nearly able to detect such motions, but the COBE measurements, being from above the atmosphere, were.

The results showed that the sun was moving with respect to the CMB at about 370 kilometers per second, which when blindly translated into a fraction of the speed of light becomes 0.001234! The direction was toward the sky constellation Leo, which, to make that direction clear to most readers, is reasonably near the direction to the Big Dipper. The CMB thus provides a universal reference frame with respect to which we can determine other motions including our own. (The motion of the spacecraft with respect to the earth and the motion of the earth with respect to the sun were removed before determining the value given here for the sun's motion with respect to the CMB; in any event, these two removed motions, together, amount to under 10 percent of the total.)

The second major discovery, and far more major at that, was the discovery of fluctuations in the CMB—that is, inhomogeneities in the temperature as a function of direction. These were discovered at a level of at most about one part in ten thousand of the CMB temperature itself and form the basis of our deductions about the formation of structure in the young universe. These COBE results have now been exceeded in sensitivity, and in frequency coverage, by those from the Wilkinson Microwave Anisotropy Probe (WMAP) and the Planck spacecraft. The WMAP was launched by NASA in 2001, and named after Dave Wilkinson, who was prominent in this research field and was involved with this spacecraft until his untimely death from cancer in 2002. The Planck spacecraft was named in honor of the early twentieth-century physicist Max Planck, who was the first to correctly characterize black-body radiation theoretically, and was launched by the European Space Agency, in 2009. Compared to COBE, the measurements from these last two spacecraft gave much higher spatial resolution as well as increased sensitivity that allowed temperature differences in different directions to be determined with an accuracy in temperature measurement of nearly a part in a million. The coldest portions of the sky run from about 200 microKelvin below the mean CMB temperature to the hottest, which are about 200 microKelvin above it. Note: 1 microKelvin is one-millionth of 1 Kelvin.

These measurements provide stringent constraints on the parameters currently used to describe the geometry and expansion of our universe. This field, although still far from having reached its peak, is now at the forefront of astrophysical research. More refined studies with more sensitive, specialized instruments are being developed and will be used to uncover more subtle properties of the universe, including the contribution of gravitational waves to the characteristics of the early universe. Gravitational waves are expected to be emitted by certain moving masses, with precise predictions following from Einstein's general theory of relativity. Until now, no such waves have been detected associated with the early universe. As of the past four or so years, however, gravitational waves have been detected from the mergers of many different pairs of black holes and of neutron stars in regions of space not nearly so distant as from the early universe. Neutron stars, by the way, are very small—only about 10 kilometers in radius!—and very dense (one thimble full has the mass of a large mountain on earth) and presumably consist mainly of neutrons. With new instrumentation being built and/or now coming into use, many scientists are hopeful that the detection of such gravitational waves from the early universe will be made in the next decade or so and provide significant new information about the history of the early universe. There is also the possibility of detecting evidence of these gravitational waves from observing certain types of fluctuations in the timing of signals from so-called millisecond pulsars, those neutron stars with spin periods of the order of a *millisecond*—that is, those that make a complete rotation in only about one thousandth or a few thousandths of one second.

In summary, we note that the discovery of the CMB places a rather firm observational foundation under the previously almost purely speculative field of cosmology. Cosmology is, in effect, the ultimate "big picture" field, the study of the origin and subsequent development of the universe on the largest spatial and temporal scales. Further, many of the follow-up observations and theoretical studies of the CMB have turned cosmology into a thriving, frontline, data-driven field of astronomy and indeed of human thought on our origins on the largest scales.

SEVEN

Dark Matter and Dark Energy

Have the previous chapters succeeded in providing the full picture of our universe? Not exactly; puzzles abound. We must for lack of space, however, skip almost all of these, such as those related to exoplanets, stars and galaxy formation and evolution, quasar and pulsar enigmas, cosmic rays, gamma ray bursts, fast radio bursts, differences in the results from different ways to estimate the Hubble constant . . . —fascinating though these all are. Instead let us concentrate on just two more topics in the first part of this book, both seemingly of fundamental importance: dark matter and dark energy.

Dark Matter

In the early 1930s, a brilliant, forward-thinking young Swiss astrophysicist began to make his mark. While studying the spectra of light from a cluster of galaxies, Fritz Zwicky noticed a peculiarity: the Doppler shifts of the spectra from the individual galaxies of the cluster spread over a rather large range. Given his estimate of the galaxy masses from their light output, Zwicky concluded that these galaxies could not be gravitationally bound to each other. Using reasonable estimates of the galaxy masses from their output of light, Zwicky inferred that the total mass was about tenfold too low for the galaxies to be gravitationally bound to one another. Of course, one might say that it is just a chance collection, and, in fact, the galaxies are

only close together for a relatively short time, sort of an accidental, tempo-
rary grouping. However, Zwicky calculated from reasonable assumptions
the probability for such a chance association and found this probability way
too small to provide a reasonable explanation of the grouping. So, what did
Zwicky propose as the explanation for this situation? Dark matter—a type
of matter not visible for some reason but sufficiently massive to gravitationally bind the galaxies together. He thought that his case was strong. What
was the reaction of the astronomical community to his paper? Nearly dead
silence. Astronomers generally thought it was interesting, but no one paid
particular heed.

Not until about forty years later did Zwicky's work capture significant attention. For example, in the 1970s Vera Rubin and her colleague Kent Ford
were measuring the light from hydrogen atoms in galaxies. (By then women
were already making a strong impact in this field.) The technology was by
that time available to determine the radial (line of sight from us) velocities of, for example, hydrogen atoms in a galaxy at different distances from
the galaxy's center. From such atoms at locations in the galaxy at different
distances from its center, they determined the corresponding speeds of the
atoms in the direction of the telescope they were using; in effect, they were
making use of the Doppler shift equation given in chapter 5. They determined these speeds for many different distances of atoms from the center
of the galaxy to its edge. The edge was determined by the decrease in the
light to a level so low that Rubin and Ford could no longer make the needed
measurements. Based on how the density of the atoms in a (pancake-shaped)
galaxy varied as their distance from the center of the galaxy increased, inferred by the brightness changes, Rubin and Ford could estimate how these
speeds could be expected to decrease as the atoms' distance from the center
of the galaxy increased. Why is such a decrease expected? Think of Kepler's third law (chapter 2) as a roughly relevant model: $P^2 = constant \times a^3$.
Thus, the average orbital speed (with v, for magnitude of velocity), v_{avge}, for
an atom in a (nearly) circular orbit about the galaxy's center, will equal the
circumference of the orbit divided by the time for an orbital period (speed
= distance/time): $v_{avge} = 2\pi a/P$, which by Kepler's third law, is proportional
to $a/a^{3/2}$, obtained from taking the square root of both sides of the third law

and substituting for P on the right side of the equation for v_{avge}. Simplifying the right side of this equation by dividing numerator and denominator each by a, we obtain v_{avge} *proportional to $a^{-\frac{1}{2}}$*, which decreases as the distance, a, of the atom from the galaxy's center increases.

The exact expression for the decrease under Newton's (or Einstein's) laws will depend on the details of the mass distribution of the galaxy; all galaxies, though, seem to have their (visible) matter heavily concentrated toward their centers. For the rest of the visible matter, we can expect the speed in our direction (the direction from the matter toward us, or the line of sight) to decrease as the distance of this material from the center of the galaxy increases.

So, given all of this overly wordy explanation, what in fact did Rubin and Ford and other scientists of that era find? Their observations did not show the materials' speed in our direction to decrease as the distance of the material from the center of the galaxy increased. Rather, as far from the galactic center as there was sufficient light for them to make their measurements, this speed remained roughly constant. It did not decrease as expected (fig. 7.1). In the illustration here, note the initial increase, at small radius (distance) from the galaxy's center, in rotation speed (approximately equal to the speed in the direction to the telescope or, equivalently, to the earth),

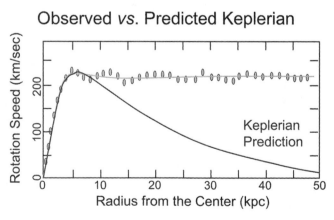

Figure 7.1. A galaxy rotation curve: observed versus predicted. Original artwork by Richard Pogge, The Ohio State University.

due to the increasing mass inside the increasing radius of the gas and stellar orbits; after a certain distance from the galaxy's center, which depends on the detailed mass distribution within the galaxy, this increase in mass would not be enough to overcome the increase in distance from the galactic center of the material, which would therefore be accompanied by a lowering of the speed in their orbits—the predicted curve. But the observed points, surrounding the black curve, didn't come close to following expectations. The *visible* light from the galaxy, however, did decrease in brightness as the observed part of the galaxy increased in distance from the galaxy's center.

One interpretation was that dark matter was present in such amounts and in such places far from the center of the galaxy that the speed of the material in our direction did not decrease as its distance from the center of the galaxy increased. Why the dark matter should be distributed such that the material's speed toward us remained (nearly) constant as its distance from the galactic center increased is *not* obvious. This type of result, coupled with the result of Zwicky's deductions from forty years earlier, caught astronomers' notice. Such rotation curves, as they are termed, were obtained for other galaxies by different groups of astronomers, all with similar results: the orbital rotation speed of the material in a galaxy did not fall off as the material's distance from the center of a galaxy increased.

Three types of explanations, or possibilities, were put forward. The first is that dark matter consists of many small objects of ordinary matter, for example free-floating planets (ones not gravitationally bound to stars) that are too small to be individually observable via ordinary light, either emitted or reflected by the objects. Hence, these objects are dark—unseen. So far no such cache of postulated objects has been found in the sky, not even black holes, which would presumably be visible by their effects on light from more distant objects. In fact, based on their observations, astronomers are fairly confident that such objects cannot be the dark matter in our galaxy, the Milky Way.

Another explanation is that dark matter is indeed some form of matter— that is, dark matter has mass, but different from what we normally observe as matter. Different types of matter have been suggested. One possible form is somewhat whimsically called WIMPs, for *weakly interacting massive particles*. They are assumed to have properties based on extensions of the current standard model of elementary particle physics. Direct evidence for the existence of WIMPs has been sought, especially on earth and in deep mines. There ultrapure materials are placed as detectors (the details here we must skip), shielded from known kinds of matter such as high-energy ordinary matter coming from outside the earth. Very sophisticated and imaginative schemes have been tried; more are being tried; others are still being developed and/or improved. Results to date: nothing detected. No hint of WIMPs yet.

A third explanation is that the problem lies with the theory of gravitation; it needs modification. Many such theories have been proposed; one, referred to as MOND (*mo*dified *N*ewtonian *d*ynamics), proposed changing the normal model of gravitation to yield the observed results from visible matter alone by, for example, increasing the effect of gravity at large distances, like the outer reaches of galaxies. So far all such proposals have led to problems in comparing some predictions with other observations.

Current status: a vast majority of scientists accepts dark matter and includes it in amounts and distributions needed to obtain a model of the universe, starting very early after the Big Bang, that leads to the structure we find today. These numerical simulations of the evolution of the universe are carried out by approximating the universe via a large number of individual particles. The highest number that can be handled now by computers is about 10^{12} or somewhat above—capability here changes rapidly! This number, large as it may seem, is *totally* insignificant compared to the estimated total number of particles in the universe. But the simulations might nonetheless give a reasonable account of the actual development of the universe.

How are the simulations done? We let the particles interact via what we currently believe are the laws of physics, in operation then as now. We start from an initial configuration that we think mimics the immediate aftermath of the Big Bang and let the system evolve under the interactions mandated by these laws. There are of course many difficulties with such arrangements. For example, we do not know the initial conditions; we cannot yet simulate all of the potentially relevant physics; and we can only employ a tiny, tiny fraction of the number of particles present in the universe. Still, such simulations are considered by most astronomers and astrophysicists to be fairly good approximations of the universe's physical evolution. The results from these simulations seem to show that dark matter comprises about 80 percent of the total matter present in our universe. (Zwicky's pioneering work in the early 1930s estimated that around 90 percent of matter in the cluster of galaxies he studied was dark matter.) More remarkably, we have literally no idea what this 80 percent is, nor, in fact, can we even be certain that it exists!

Dark Energy

Since the universe seems to be uniformly expanding, what can we expect for the future? Will it keep on expanding at its present rate, described by the Hubble constant, H_o? (The subscript "o" signifies its present value, which is still somewhat controversial.) Or would we expect the rate of expansion to change? If so, would we expect it to increase or to decrease as time progresses?

According to Newton and, in effect, to Einstein, too, matter attracts other matter, leading us to expect the universe's matter to attract itself to itself and therefore its expansion to slow down. This situation parallels a more familiar one: we throw a ball up from the surface of the earth; it rises, but slows down as it rises, independent of any friction from the presence of the atmosphere. In the universe's case, each part of it is expanding away from all of the other parts; because of their masses, all parts also attract all other parts, so that their expansion rate would be expected to slow down. We can estimate for our present universe what we would expect this slow-down of the rate of expansion, or deceleration, to be. It turns out to be very small, requiring an experimental tour de force to detect. How could we do it? We could look at objects both nearby and very distant, and we could look for a change in the rate of expansion of objects away from us (their recessional velocity)—that is, a change in the Hubble constant—as shown by wavelength shifts in the spectral lines of the objects.

There are several possible scenarios for the expansion rate of the universe over time (fig. 7.2). In the illustration here, at each point of each curve,

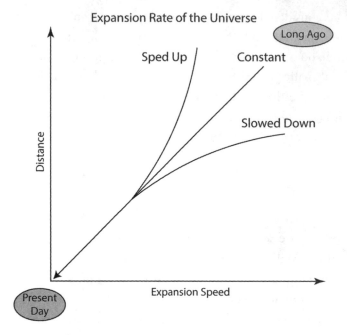

Figure 7.2. Possible scenarios for the history of the expansion rate of the universe from long ago to the present time (see text). Courtesy of David Shapiro.

the slope—the change in distance per unit change in recessional velocity (or expansion speed)—gives the inverse of the rate of expansion, in effect the inverse of the Hubble constant: the larger the slope, the smaller the Hubble constant (and the lower the expansion rate), and vice versa. If the recessional velocity was in the past less than expected for a constant rate of expansion (the topmost curving line), then the expansion rate was lower in the past than in the present. The lowermost curving line shows the opposite possibility, that the expansion rate was greater in the past, and has slowed down—the Hubble constant has decreased over time.

To recapitulate: if we were to determine the rate of expansion, H, by measuring distances and velocities of *faraway* galaxies—that is, to look farther back in time—we would expect to find the value of H to be larger than when we made such measurements of *nearby* galaxies. Why? Because this would mean that as the universe is expanding, the rate of expansion is slowing down. This slowdown is expected since the mutual attraction of the matter would be expected to slow the rate at which the pieces of matter separate from one another. Thus, we would expect to find H to be greater the farther back in the past—the farther out in space—that we look.

Scientists decided to try to detect this expected (very small) change in H by using the most precise, and presumably most accurate, long-distance measurement tool in their box (see chapter 5), which occupies the top rung of our cosmic distance ladder, the particular type of supernova, SN 1a. To make this measurement, two main groups competed, one centered at UC Berkeley (the Supernova Cosmology Project), the other at the then Harvard-Smithsonian Center for Astrophysics (the High-Z Supernova Search). (Z symbolizes a quantity that relates to distances of objects from us: the greater the distance, the larger the value of z.) The rivalry between these two groups was intense. The end result was as startling as it was spectacular: the data showed that the Hubble constant was not *larger* in the past (less than five billion years ago) but *smaller* (fig. 7.3). The Hubble constant is *increasing* with time! In other words, the universe is accelerating, not

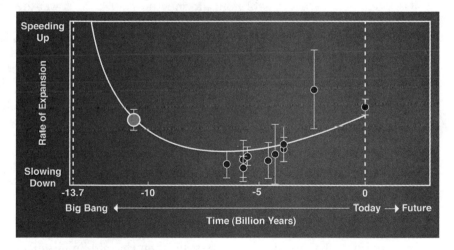

Figure 7.3. Sample measurements of the relative rate of expansion of the universe from about eleven billion years ago until today. Note that the initial slowing of the rate of expansion was overcome about five billion years ago, after the density of the universe had decreased sufficiently. (Note: The density, the amount of mass-energy per unit volume, decreases as the universe expands because we have the same amount of mass-energy [recall $E = mc^2$, Einstein's famous formula relating mass and energy] occupying a larger space.) © 2010 The Regents of the University of California, through the Lawrence Berkeley National Laboratory.

decelerating, in its expansion. There exists, in effect, an antigravity force of some sort pushing galaxies apart, overcoming and surpassing the attractive force we all know and love, that pulls matter (and galaxies) together. This antigravity force did not cause an acceleration when the universe was very young and very dense; under those conditions the strength of the attractive gravitational force was large enough to keep the expansion of the universe decelerating. But about five billion years ago (see fig. 7.3), as the universe got big enough for the density, and hence the gravitational forces, and their general relativity analogs, to become small enough, antigravity won out. As a result, the change in the expansion rate became positive — that is, this change caused the expansion of the universe to accelerate, as we now observe.

Was this quite startling result simply accepted by the relevant scientific community? Not quite. All sorts of potential problems with the use of SN 1a's as distance indicators had to first be investigated carefully. A major example is the uncertainty in the amount of dust along each line of sight. By absorption of some of the light from the supernova, dust can lower the apparent brightness and hence increase the inferred distance to the SN 1a. Another such potential problem was differences in the intrinsic brightness among the supernovas that we lumped into this one class. Each potential problem was considered and found to be either correctable or too small to affect the conclusion on the accelerating universe. Thus most scientists now believe that the expansion of the universe is accelerating; the energy associated with this acceleration is called dark energy. For this discovery of the now accelerating universe, there were three resultant winners of the Nobel Prize, from the two collaborations of many people each, alphabetically: Saul Perlmutter, Adam Riess, and Brian Schmidt. The second and third did their major work leading to this discovery of the acceleration of the expansion of the universe while they were at the then Harvard-Smithsonian Center for Astrophysics; the first was at the University of California Berkeley and had been a Harvard undergraduate. Two and a half cheers for Harvard!

What is underlying this startling aspect of nature's behavior? No one knows. Physicists find that the standard model of elementary particle physics allows for a negative pressure in the vacuum. (Don't let this seeming oxymoron bother you, however; we don't have space to explain this issue in detail even if we did understand it.) But some calculations indicate that it's either exactly zero or approximately 10^{120} times larger than observed—a mind-boggling contrast. What scientists have also realized, however, is that they can resurrect Einstein's cosmological constant from the dustbin of history and fit well all of the supernova data gathered so far on the increase in the rate of expansion of the universe. Not that such provides any further understanding. But it does give us a model that seems to work.

To put a finer point on this situation: the cosmological constant is totally ad hoc; it doesn't follow from any known principle of physics. Many scientists think that this newly recognized accelerating universe seems to be implying that we are in need of a new, dramatic insight into physics and the structure of the universe. Some scientists hope that a new model breakthrough will be as important to the twenty-first century as the quantum and gravitational breakthroughs were to the twentieth century. Most scientists doubt such an outcome. Time will tell. Meanwhile, if we take stock of the total mass and energy involved in the accelerating universe, we find that the so-called dark energy contributes about 70 percent of the total, with dark matter contributing about 25 percent and our ordinary mass and energy the remaining 5 percent. In this accounting, matter represents about 30 percent of the total, with about 85 percent of the matter being dark and about 15 percent being the matter with which we are intimately familiar, as noted earlier. Thus, we do not understand in the way that we'd wish the origin or even the existence of about 95 percent of the universe—if you like, a further downgrade of humans from their earlier arrogation of a key place at its very center.

There are three more questions that I wish to treat briefly before ending this first part of the book: Why are there more particles than antiparticles? Why is the gravitational force so much weaker than the electromagnetic force? And last, is there more than one universe?

Why Are There More Particles Than Antiparticles?

As scientists now know, for every charged particle, for example, electron or proton, there is an antiparticle, an identical particle except for the sign of its charge. For example, in 1932 scientists discovered the positron, the antiparticle of the electron (see chapter 11). An amazing property of anti-particles is that when one collides with its particle counterpart, they destroy each other while maintaining energy and momentum conservation—two of the fundamental laws of physics discovered well over a century ago. The products of these collisions are photons, which have total energy and momentum that match those values for the particle and antiparticle before the collision. How come the universe around us seems to consist overwhelmingly of particles rather than of antiparticles? This condition is abundantly clear on the earth, albeit less so in more distant parts of the universe. How did this condition arise? What led to this asymmetry? This line of inquiry is clearly fundamental. Equally clearly, no one yet has a good clue as to the answer. This is the way nature seems to behave, but we do not yet have a good, testable model for this behavior.

Why Is the Gravitational Force So Much Weaker Than the Electrical Force?

We know two macroscopic forces that each act over long distances: electrical and gravitational. How do they compare? Particles carrying charge—not all do—come in two flavors, called positive and negative, with those of the same sign repelling and those of opposite sign attracting one another. By contrast, except for the largest spatial scales, explored in this chapter, under gravitation all particles attract one another. In gravitation, mass is the analog of charge. Mass comes in only one flavor, though, not two as for charge. So, instead of having some particles attracting one another and some repelling, as for charge, with mass, all particles attract one another. Mysterious. We have no good model yet for these observed properties of nature.

What about a comparison of the strength of these two types of forces, electrical and gravitational? An apparently sensible way to make this comparison is to take, say, an electron and a proton and compare the strength of electrical and gravitational forces as they affect these two particles. Both forces, to a very good approximation, follow the inverse square law for the effect on them of distance of separation. Upon doing the calculation, we find that the electrical is nearly 10^{40} times as strong as the gravitational force.

Why is there such a whopping difference? It's an interesting question; we have no good model; it's just the way nature behaves. One more question: Why then are gravitational effects of any consequence at all compared to electrical ones? Answer: because large collections of particles are very close to electrically neutral and so the electrical forces tend to cancel out. Not so gravitational forces, which have only one sign of charge—mass is always positive (by convention). So, with very large numbers of particles, gravitational forces must be reckoned with and usually dominate.

Is There More Than One Universe?

Why is it that the speed of light has the particular value that it has? Why is it that the electron has the specific charge that it has? Why does the gravitational force have just the value for Newton's gravitational constant that it has? A similar question can be asked about each of the other constants of nature. One possible answer is simply that that's the way nature is. Another of many possible approaches is to say that we are just in one of a huge number of separate universes, each one of which has different values for these constants. There are also other theoretical reasons for thinking that many universes might exist (the concept of multiverses), but we dodge a detailed examination. Let us merely note that some theorists estimate the number of such universes to be the order of 10^{500}, an absolutely mind-numbing number. There is now no known way to test this multiverse hypothesis. So, we are here perhaps now more in the realm of science fiction than of science: the theories are not known to be falsifiable by probing nature. As for the future . . .

The Earth and Its Fossils

EIGHT

The Origin, Shape, and Size of the Earth

From whence the earth? This question has intrigued people for millennia. Our model of the origin of the universe and the development of structure within it notwithstanding, we still do not have a convincing, detailed model of the origin of planets.

In broad outline, scientists have conceived of the following: inhomogeneities in the products of the Big Bang led to collapses under gravitational attraction, thus forming first-generation stars. These (atomic) products are limited, as far as we know, by the Big Bang and the laws of physics to be primarily hydrogen, helium, and a small amount of lithium. These first-generation stars eventually ran out of their nuclear fuel; many of them then collapsed and, remarkably, exploded. The explosions released into interstellar space all of the heavier atoms, or elements, that they had manufactured from their nuclear fuel; many heavy elements were produced by colliding neutron stars. There, they formed into gas and dust (rather small solid particles), the latter presumably via a concatenation of collisions of gas atoms and molecules, which presumably formed from other atomic collisions. Due to inhomogeneities in the distribution of this material in space, its space configuration tends to be unstable and to gravitationally collapse, much of the material over time turning into second-generation stars. The mass would be large enough, as with first-generation stars, for the shape of the star to become roughly spherical, under its components' mutual gravitational attraction. Some of this material, such as that part which has sufficiently high velocities relative to the others, avoids this collapse and is left orbiting the now central star, usually in a roughly pancake shape. This orbiting material is also composed of gas and dust, with both the gas and the dust particles tending with time to collide with one another. Since at a collision the particles are orbiting at the same distance from the central location (the star), they usually collide at relatively low mutual velocities and thereby often stick together during these collisions. Whether two objects upon colliding stick together or break into smaller pieces, or partly each, likely depends upon their relative velocities, orientation at impact, shapes, sizes, physical structures, and compositions; in brief: it's all likely very complicated.

The above paragraph is wholly qualitative; it is not easy at our present state of knowledge to be very quantitative and reliably so. Thus, we qualitatively conclude that via collisions this leftover dust and gas coalesce into larger objects, which we label as planets when they get suitably large. The details here, despite having been simulated in numerical experiments with large numbers of particles, do not seem to follow from the qualitative description just given. There apparently is a difficulty getting to sizes larger than some meters in diameter via this collision mechanism. Clearly, of course, planets do form, so nature has managed to surmount this problem found with our simulations. Understanding how nature has accomplished this task is a forefront area of research with a variety of opinions contending for supremacy. In other words, the dust has yet to settle on this issue (pun intended).

After forming, we believe that the earth was at least partially molten, particularly in the center. The lower down material is, the more material that overlays it and presses down on it, and the higher the pressure and the higher the temperature as we go toward the earth's center. Denser melted materials, such as iron, sunk toward the center. Somehow, it's not yet clear exactly how, a lot of water appeared on the surface, with gases above forming the atmosphere—voilà! Our earth. Our earth has substantially evolved over time, especially its atmosphere and surface, as we'll note later on. But one fact we should always keep in mind: the earth and we humans—as well as all other organisms!—are made, except for our hydrogen, mostly from atoms that were created in stars.

Now that we have a tiny glimmer of how the earth might have formed, we move on to discuss its properties.

The Shape of the Earth

How did people millennia ago figure out the shape of the body they lived and walked on? It wasn't obvious. Today when people take extended walks or travel over most parts of the earth, the ground appears to be very uneven, having what we call hills, mountains, and valleys disported over its surface. All well and good, but what is the overall shape of this body? It is not known when humans first recognized this shape. It is, however, a good bet that some ancients were aware that this shape was roughly spherical. How would they have reached this conclusion? By looking up, for example, and noting that during eclipses in which the earth is between the sun and the moon, the earth's shadow on the moon looks circular. Knowledge of geometry was most likely sufficiently advanced that the intelligentsia of that era realized that the projection of a sphere onto a plane was a circle. Although the moon is not a plane, we are talking here about the plane of the sky. The fact that the shadow of the earth moves across the face of the moon during an eclipse always appearing as part of a circle likely led inexorably to the conclusion that the earth must be, at least roughly, in the shape of a sphere (fig. 8.1).

There are other ways to determine that the surface of our earth is at least curved. The classic example is a ship sailing from a port. As it recedes from view, regardless of its direction of travel, the first part of it that disappears is the bottom and the last part the top. The obvious interpretation of this sequence is that the earth is curved. While this evidence, by itself, is not conclusive, the idea that the earth is (close to) spherical is certainly *consistent* with these facts. In recent times, especially after the start of the Space Age, with pictures of the earth taken from space, and also following circumnavigations of the globe nearly half a millennium earlier, the evidence for the earth's shape has been more direct and rather convincing.

Figure 8.1. A modern photo of a partial eclipse of the moon by the earth. The circular aspect of the shadow cast by the earth should be clear. Photo by Aerialpete, CC BY-SA 2.0.

The Size of the Earth

Having discussed some of the various ways through time that led to the inescapable conclusion on the earth's shape—save for an escaped minuscule minority of people—we now confront the issue of the earth's size. Even to the ancients, it must have been quite clear that the earth was very large compared to the size of humans. But how large? Those scholars who have looked at the history of our knowledge of the size of the earth usually credit the first solidly based, quantitative estimate to have been made by Eratosthenes in about 250 BCE. Before delving further, I remark that no original work of Eratosthenes has apparently survived, and much of the following history might be unreliable. Such is the difficulty in dealing with the distant past. What do we think that Eratosthenes did? He apparently believed that the earth was spherical and set out to measure its circumference (fig. 8.2). He had the super idea that if he could measure the angle, θ, between two radii from the center of the earth and measure the distance between these two radii along the surface of the earth (arc-length, D), then the circumference would follow by simple proportion: the angle, θ (in degrees), divided by 360 equals the arclength, D, divided by the circumference, C, of the earth, in the same units as used for the arclength. Thus, we can by simple algebra solve for the circumference, C:

$$C = D \times 360/[\theta \text{ (in degrees)}]. \tag{8.1}$$

But how do we measure D and θ? Let's first set the stage by discussing the choice of the end points of the arc, D. The first point is easy: Eratosthenes lived in Alexandria, Egypt. The second point was chosen based on an interesting happenstance. It was well known that there was a water well on the island of Syene in the Nile River, far south of Alexandria, where each year on the summer solstice, June 21, at noon sunlight shone directly down the well. A single straight line thus passed at that instant from the sun through the surface of the well to the center of the earth (fig. 8.2). We can also see that Eratosthenes in Alexandria could, at noon on this day, measure θ as the angle between the vertical and the direction of the sun. I assume that he was

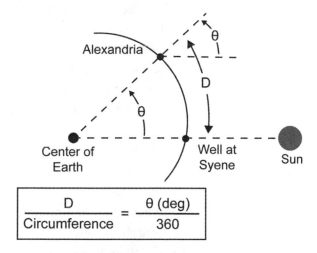

Figure 8.2. The geometry used by Eratosthenes to measure the circumference of the earth. Note: the distance to and size of the sun are obviously *not* to scale. Courtesy of David Shapiro.

likely able to measure this angle with an uncertainty of only a few tenths of 1 degree with equipment then available to him.

What about measuring D, the distance between Alexandria and Syene? We assume that Eratosthenes used a unit of distance called the *stadion* (plural *stadia*), which was then in common use, but we don't know exactly how he made his measurement. Speculations abound. In one, it is said because the land in Egypt was well surveyed at that time, all that Eratosthenes had to do was consult the survey results to determine the needed distance. Another speculation is that it was "well-known" how long camels took to walk from Syene to Alexandria, and the reverse, and how fast, on average, camels walked over that distance. With these figures in hand, simple arithmetic sufficed to estimate the distance between Alexandria and Syene. Although I have not seen the original evidence, I understand that Eratosthenes's result for the circumference of the earth was about 250,000 (± 50,000) stadia, with the angle, θ, having been measured to be 7.2 degrees (one-fiftieth of a circle), and with D = 5,000 (± 1,000) stadia. (The value 5,000 is likely an approximate number; it seems to me very unlikely that, for example, putting together land surveys over such a large distance would yield a result of exactly 5,000 stadia.) How good was this estimate, compared to our present value? We've thus come upon the last problem in this determination: What was the length of a stadion in modern distance units, such as meters and kilometers?

For those with a historical bent, this question is fascinating to investigate. The stadion was apparently defined as the length of the path surrounding the original Olympic stadium in Athens. Since that stadium has been rebuilt twice between then and now, since no one knows exactly what path around it defined this unit of distance, and since no one seems to know how or how accurately that distance was transported to Alexandria, there is clearly room for argument. My considered (amateur) conclusion, based on my readings, is that the stadion was equivalent to about 175 meters, with an uncertainty of about 10 percent. Assuming as above that the estimate of the distance, D, was independently uncertain by twice this percentage, we conclude that in modern distance units, Eratosthenes's determination of the circumference of the earth was about

$$C = 44{,}000 \pm 10{,}000 \text{ kilometers.} \tag{8.2}$$

What made Eratosthenes famous, though, was the *method* he invented rather than the *result* he obtained. All aspects considered, this result compares well with our present knowledge of the earth's circumference, which in the North-South direction that Eratosthenes used is about

$$C = 40{,}000 \text{ kilometers.} \tag{8.3}$$

Since Eratosthenes's time, measurements of the size of the earth have continually improved, if rather sporadically. As we noted in chapter 4, for example, Jean Picard in about 1670 measured the length of a degree of latitude on the earth's surface, obtaining a result within about 0.4 percent of the modern value (and, according to one apocryphal story, resulted in Newton's obtaining agreement with the earth's causing the moon and an object near the earth's surface to both be accelerated in accord with his model of gravitation).

Thanks to space geodesy—using space to make accurate measurements of locations of places on the earth, a subject treated in chapter 10—we can now measure the distance between many places on the earth even a continent or more apart with an uncertainty at the millimeter level; we also can measure the mean circumference of the earth in the polar direction to about one part in 10^9. In addition, we can measure the distance from the center of the earth to the surface at many places and many times to centimeter accuracy. Why do I write, "many times"? Because of the fact that the earth's radius at any given location changes with time! For example, there is a seasonal effect due to movement of water and there are diurnal and semidiurnal tides of several tens of centimeters in magnitude. That we can monitor such changes represents an astonishing improvement in measurement accuracy by a factor of about one thousand to ten thousand in somewhat over half a century, the present extent of the discipline of space geodesy.

With the shape and size of the earth under our belts, we move now to examine its insides.

NINE

The Earth's Innards

How can we learn about the internal structure of the earth? We can't just drill to the center. Even with our most advanced equipment, we've only been able to drill in one place down to about 12 kilometers, barely 0.2 percent of the way to the center, merely scratching the surface so to say. (Drilling down much farther, though, seems all but impossible since all human structures made so far would not survive the pressures at much lower depths but would instead collapse.) Still, drilling down 12 kilometers is hardly an unimportant accomplishment, given what we have learned, and can still learn, about, for example, past climate from the material recovered. Drilling down is in part probing the past, somewhat akin to looking out into space and thus back in time.

What could we learn before about the earth's insides? Over two centuries ago, we could calculate two overall properties of the earth: its mass and its average density (mass per unit volume). The former we get from measuring an acceleration of a falling object at the earth's surface and using Newton's laws of gravity and motion, along with values for the radius of the earth and for Newton's constant of gravitation, G. The result of this measurement (about 10 meters per second per second) and computation, using the measured radius of the earth and the measurement of G, is about 4×10^{27} grams, or about 10^{25} pounds. To calculate density, we use this mass and the volume of the earth, which follows from its shape and size. The value we obtain is a quite hefty density of slightly over 5 grams/centimeter^3, more than five times the density of water. This result shows that the matter in the interior of the earth is more tightly packed than at the surface, since rocks near the earth's surface have a density of only about half this value, approximately 2.5 grams per cubic centimeter. These are overall, global properties. What about the details of the earth's internal structure?

As for the Cosmic Distance Ladder, we use mostly proxies to learn about the earth's interior: specifically, we use earthquakes. This fact is somewhat ironic, since one of the main reasons why we want to study the earth's interior is to be able to predict earthquakes. This goal still eludes our intensive efforts but is clearly worth the chase.

Seismology

Earthquakes inspired some of the oldest comments we know about natural phenomena, starting back almost four thousand years ago. One earthquake was recorded in 1831 BCE in Shandong province in northeastern China. The great devastation sometimes caused by these sudden, unpredicted phenomena obviously caught people's attention! Aristotle, as usual, also had his say. Apparently, his theory was that earthquakes were caused by winds within the earth. The first example of what we would consider a scientific theory is attributed to an Englishman, the Reverend John Michell, an eighteenth-century polymath. Michell reasoned that earthquakes were due to "waves set up by shifting masses of rocks miles below the [earth's] surface"—a very perceptive description. Michell's interest was likely sparked, or was renewed, by the truly calamitous earthquake that struck Lisbon, the largest city in—and capital of—Portugal, in November 1755. The origin of this earthquake was located about two hundred kilometers west-southwest of the southwest tip of Portugal and was felt all over Europe and even beyond.

So how can we use earthquakes as a proxy for probing the earth's interior? An earthquake is characterized by displacement with respect to their neighbors of small gobs of rock matter relatively near the earth's surface. Such displacements excite waves in the earth's materials. These waves are of different types, each having its unique characteristics. Detecting these waves as they impinge on the earth's surface is the first step in their study and interpretation. The ensemble of these processes of detection and study forms the subject matter of seismology. This subject gives us most of our information about the structure of the earth's interior. We thus now discuss the basics of this subject and some of the main results so far obtained.

Instruments used to detect and record the motions of the earth's surface, mostly due to earthquakes, are usually called seismometers. A precursor instrument from China, the seismoscope, gave only the direction from the earthquake. Though we don't know what it looked like or exactly how it worked, it seems that the mechanism involved balls that sat in dragon-shaped cups attached to its sides; when an earthquake tipped the apparatus, the ball on the side that tipped down would be dislodged, indicating the direction of the earthquake. Rather than dwell on this seismoscope, let us skip to a schematic of a modern seismometer to explain the basic principles on which it operates (fig. 9.1).

The seismometer has a frame. It must be rigidly attached to the ground so that it partakes of any vertical motion the ground might exhibit due to the effect of seismic waves passing through that part of the earth. When

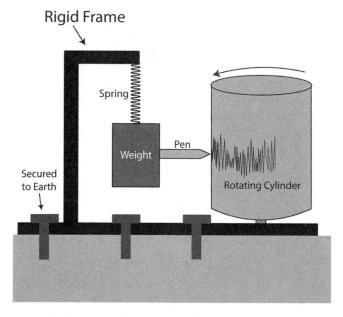

Figure 9.1. Schematic of a simple seismometer sensitive to vertical motions. Courtesy of David Shapiro.

the frame moves up and down, the weight hanging from the spring, being nearly free from connection to the ground, will not move up and down rapidly with the frame. The pen, rigidly attached to the weight, and pressing against the rotating drum, which is firmly attached to the frame, will record the vertical motions of the weight with respect to the frame, preserving a record of the motion of the frame. A key question is how does one relate the recording on the drum to the vertical motion of the earth at this location? Determining this relation is called calibrating the instrument. One may put the frame, for example, on a platform, which is isolated through absorbing material from the earth below, and then move the frame up and down at an accurately determined rate and amplitude. One may do this for a wide variety of rates and amplitudes to mimic those that will affect the frame during various kinds of earthquakes. Knowing these relations for the frame would allow scientists to interpret the curves gathered by the rotating drum during earthquakes. Actual calibration is done in different ways; the method outlined here is intended only as a proof of principle.

What about horizontal motions of the earth? Detection of these is accomplished in a similar, if somewhat more complicated manner. For example, there are two dimensions one has to deal with horizontally—north-south and east-west—instead of just one, up-down. But the basic principles are the same.

We also want to know the time at which the motion recordings are made, for example to compare recordings made at different locations. Thus, a seismometer is coupled to a more than sufficiently accurate clock, which is available and ensures that the recording of the motions is properly connected to the time. Such recordings are now made fully digitally.

What are these earthquake motions like, and how can we describe and interpret them? Here I will be a bit didactic—telling you first what the waves are like before showing examples of the evidence. Seismic waves—waves excited in the earth by movements of material there, as in earthquakes—come in three basic varieties: body waves (traveling waves), whole-earth oscillations (normal modes), and surface waves. Let us discuss only the first two types—we can't do everything. The body waves are of two main types, P and S. "P" stands for pressure, push-pull, or primary waves; "S" stands for shear, sideways, or secondary waves. Pressure waves travel through the earth

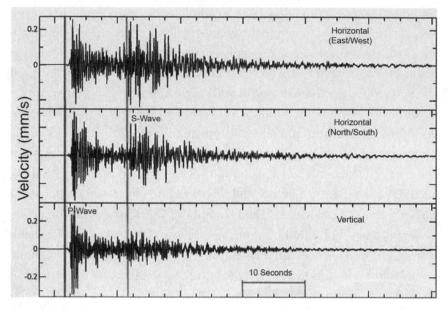

Figure 9.2. P and S waves recorded by a seismometer. Note the slight motion that exhibits the arrival of the P wave just to the right of the leftmost vertical line. The arrival of the S wave is not so obvious from this figure alone. Wikimedia Commons, User: Crickett.

at speeds greater than those at which shear waves travel. Hence, pressure waves will arrive at a given seismometer before the corresponding shear waves, excited by the same motion of earth material, as in an earthquake. The terminology of "primary" and "secondary" also denotes the relative arrival times of these two types of waves (fig. 9.2). (Note that the vertical scale for the middle plot should be labeled the same as in the top and bottom plots.) The recording of the P wave shown here started about ten seconds earlier than that of the S wave, indicating that the P waves arrive first at points away from the earthquake origin.

How do these waves propagate (move through a medium)? For P waves the motion of the material is in the same direction (or opposite to that direction) as the propagation. By contrast, the key point is that the material motion for S waves is sideways, the motion being perpendicular to the direction of propagation of the wave. Thus, we have some of the meanings for P and S waves: push-pull and sideways, respectively.

For a P wave traveling in a rock, the rock particles move closer together, then farther apart, then closer together . . . in the direction that the wave is moving, as time increases at any part of the rock. Correspondingly, for a traveling S wave, moving left to right, the particles could move up and down at this part of the page, as time increases. Or, for example, the particles could move in and out of the rock or in any other direction, perpendicular to the direction of motion of the wave.

Inferences about Earth Structure

We now consider the paths that P and S waves take when traveling through the earth (fig. 9.3). In the cross-sections of the earth shown here, the propagation of P waves (black lines) continues through the core of the earth, while S waves (also black lines) move only through the earth's solid mantle. Why is this? Because *shear waves do not propagate through a liquid*. Sideways motion of a particle in a liquid does not affect its neighbors in the direction of propagation, so when a shear wave encounters a liquid, the wave stops propagating. In the figure, the circular band near the center of each of the two parts signifies, at least at its outer edge, a liquid. Through this pattern of (non)propagation, scientists inferred the existence of a liquid core in the earth.

There is one other striking feature of the figure: the path of each wave is *curved*, leading from the point where the waves are initiated at the top of the drawing to where they again hit the surface lower down. These lines are curved basically because the density of the material of the earth is different in different locations, here increasing as we proceed to lower depths in the mantle. Other properties and the composition of the material also change as we go from the surface to the center. These changes cause corresponding changes in the direction of the propagation when the propagation is not radial. For this same reason, when one shines a flashlight into water in a direction not perpendicular to the surface of the water, the direction of the propagation of the light changes. In particular, as the light passes from the air into the water, the light bends toward the direction perpendicular to the surface.

While no S waves are drawn in almost all of the bottom half, they do appear there. This penetration happens in two ways. The first: S waves move

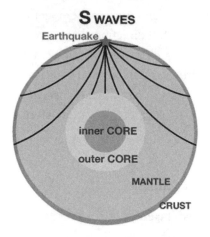

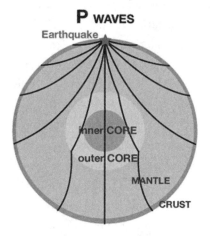

Figure 9.3. S and P waves propagating through the earth. U.S. Geological Survey, Department of the Interior/USGS.

mostly sideways and travel only a relatively short distance before intersecting the surface; although not shown here, the waves are partly reflected and move then at reduced strength another short way before again intersecting the surface, and so on around the earth, until they peter out, having lost virtually all of their strength. The second: when a P wave intersects the boundary between the outer and inner core, the P wave is converted in part into an S wave—this is how nature behaves! This S wave in the inner core propagates until it again reaches the outer core, at which point it is partly converted to a P wave. This P wave then propagates serenely until it strikes the boundary with the lower mantle. At this intersection, the P wave is again converted in part to an S wave. This S wave keeps on propagating until it hits the surface, and so forth. Complicated paths are possible and are inferred from recordings at seismometers located at multiple sites on the earth's surface.

So what can we learn about the structure of the earth from all of these recordings? From each of many earthquakes in various parts of the globe, one can monitor the times of arrival and the forms of the various waves that strike each of a globally distributed set of seismometers. One can infer, for example, the velocities of propagation of both the P waves and the S waves as a function of depth within the earth. The velocity varies primarily with depth, although there is certainly heterogeneity that causes the velocities to vary as well with the other two directions. The main variation, however, is with depth (fig. 9.4).

Here we see that from a depth of about 2,900 kilometers to a depth of nearly 5,200 kilometers the speed of the S waves vanishes, main evidence for the presence of the liquid core. Below this liquid core is a solid core, discovered in about 1935 via clever scientific detective work, using in part the techniques noted above, by the Danish geophysicist Inge Lehmann; she lived to the age of 104, one of the longest-lived among scientists. We also see the rather large difference between the speeds of the P waves and the S waves, which grows with depth and reaches about 8 kilometers per second in the solid inner core. Of course, the difference is even larger in almost all of the outer core, where the S waves vanish.

We can infer from these and other data the approximately spherical, layered structure of the earth. There are also fine-grained results from relatively small structures not distributed through the earth in a spherically symmetric manner. We will discuss some of these in a later chapter.

Earthquake Location

What can we learn about earthquake characteristics from monitoring, at a network of seismic stations, the waves initiated at the earthquake? By

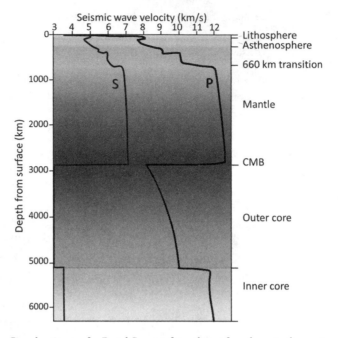

Figure 9.4. Speed estimates for P and S waves from the surface down to the center of the earth, with the uncertainty generally increasing toward the center (except for the S wave speed in the outer core!). Note, too, the sharp decrease in the P wave speed at the boundary between the mantle and the outer core. Adapted from data published in Stephen Earle, *Physical Geology*, 2nd ed. (Victoria, B.C.: BCcampus Open Education, 2019), 300. CC BY 4.0.

comparing the times of arrivals of the P and S waves at the various stations, one can infer by triangulation the location of the earthquake's epicenter (the point on the earth's surface directly above the earthquake). This is easy to say, but how do we realize it in practice? Let's consider first just one station. We look at the seismometer for the P wave and for the S wave, noting the difference in the first time of arrival at the seismometer of each wave. Since we now know the difference in the propagation speeds as a function of depth of these two types of waves, if the earthquake took place relatively nearby, so we could safely use the near-surface velocity for each of the P and S waves, we can estimate reasonably well the distance of the epicenter from the seismometer. We can also for pedagogical purposes neglect the curvature of the earth. Under these conditions and assumptions, how do we estimate the location of the epicenter? Knowing the speed of travel of each of the two types of waves, we can use the difference in travel times—the difference in first times of arrival of the two wave types—to determine the total distance, d, the waves have traveled: the total time the shear waves traveled is the distance they traveled divided by their speed of travel (time equals distance divided by speed). The same goes for the total time the pressure waves traveled. The difference between these two travel times is the difference in the time of travel, or equivalently, the difference in the times of arrival at the seismometer of the two types of waves. So, in our equation, below, the only unknown is the distance, d, for which we can solve using algebra:

$$(d/v_s) - (d/v_p) = t_s - t_p, \qquad (9.1)$$

$$d = (t_s - t_p)/([1/v_s] - [1/v_p]), \qquad (9.2)$$

$$= (t_s - t_p)v_s v_p/(v_p - v_s), \qquad (9.3)$$

where the definition of the other symbols should be clear; for example, v_p is the speed of the pressure waves.

Knowing the distance from the seismometer to the epicenter, we know that the epicenter is located somewhere on a circle, centered at the seismometer with radius equal to the distance to the epicenter. But where on the circumference of the circle is this epicenter? With only one seismometer taking data from the earthquake, we cannot tell. If we had another seismometer located elsewhere and took the same data, we could determine that the earthquake lay on the circumference of a circle centered at the second seismometer's location. We can then expect that the two circles will intersect, in general, at two different points, both lying on each of the two circumferences. We have therefore reduced the possible location of the epicenter to a choice between two locations. To distinguish between these two, we use data from a third seismometer located at a point different from the first two and not collinear with them (why are these caveats crucial?). The intersection of all three circles will then be a single point (fig. 9.5). Via this long-winded description, we have shown how to locate approximately the epicenter of an earthquake!

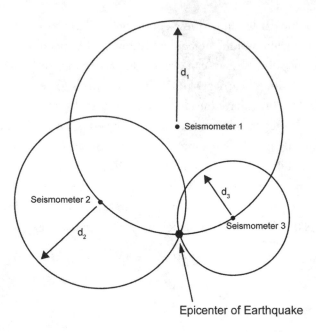

Epicenter of Earthquake

Figure 9.5. Triangulation to locate the epicenter of an earthquake from difference of times of arrival of S and P waves at each of three (noncollinear) seismometers. Courtesy of David Shapiro.

The Richter Scale

How do we tell the strength of an earthquake? More precisely, how do we determine the amplitude of, or the energy released in, an earthquake? A scale, developed in 1935 by Charles Richter, is unsurprisingly enough called the Richter scale but is somewhat shaky (pun intended)—that is, it is not a very accurate measure in application. The amplitude is given in the scale as a *magnitude*, namely as the logarithm of the maximum amplitude compared to some appropriate standard. To estimate the maximum amplitude of the motion at the earthquake, one needs to know, for example, where the location of the earthquake is relative to the location of the seismometers, the amplitudes of ground motion measured by the seismometers, and the characteristics of the medium through which the seismic waves passed. The relation of the energy released by an earthquake to the maximum amplitude of the earthquake motions is also not straightforward. These complications aside, we can say that the energy released in an earthquake is about thirty times greater for an earthquake one magnitude higher than another on the Richter scale. To give you a feeling for these magnitudes, note that an earthquake of magnitude 2 is barely felt by humans, even if they are located very near to its epicenter. On the other hand, a magnitude 9 earthquake, such as the once-in-a-century earthquake that struck in Chile in 1960, is *enormous* in its impact.

The Richter scale has been largely replaced over the past two decades in scientific circles by a scale based on an (arcane) measure of the "seismic moment" of an earthquake. We will not pursue this newer scale further, but I point it out, since you may well see it occasionally mentioned in newspapers in articles on earthquakes.

The Temperature of the Earth

Before leaving the earth's interior, I feel compelled to say something about its temperature. How do we take it, to know the extent of its fever? Seismology is not directly of great help. But knowing the earth's structure from seismological probing does provide indirect assistance. Thus, from the density of the material as a function of depth and from its state (for example, the liquid outer core), we can make useful inferences about the composition. From laboratory studies, at high pressures and temperatures, scientists have inferred that the inner core must be composed primarily of iron and nickel — heavy and dense elements present during the formation of the earth and probably sinking to the core when the earth first melted, in the later stages of its formation. The density of the liquid core matches closely that of a mixture of these elements when subjected in the laboratory to pressures comparable to that of the core.

By such means we can extend from our direct measurements of temperature near the surface and somewhat below, for example, in mines, to temperatures down to the earth's center. The results for temperatures deep within the earth, however, cannot be considered reliable because of the various (unreliable) inferences made in the estimation process. That caveat having been given, I direct your attention to a display of near-current estimates of temperature versus depth in the earth (table 9.1). For this somewhat crude — and correspondingly inaccurate — listing, one can consider the temperature to increase linearly with depth between entries, which

Table 9.1 The approximate increase of
temperature with depth within the earth

Depth (km)	Temperature (K)
410	500
660	1,900
2,900	3,000
5,150	5,000
6,370	7,000

correspond to features (for example, the boundary of the mantle and the outer core).

Here is the bottom line: the temperature becomes very high very rapidly, reaching values of the order of 7,000 Kelvin near the center—comparable to the sun's *surface* temperature, its central temperature being approximately 15 *million* Kelvin.

Fracking, Earthquakes, and Contaminated Water

Let us end this chapter by exploring a recent social problem involving earthquakes in the United States that illustrates how science can be strongly, and controversially, involved in social problems.

First, a bit of background. Until about fifteen to twenty-five years ago, earthquakes in the Midwest were recorded to be very rare and of low magnitude. Then they increased, both in frequency and magnitude, and dramatically so in the past decade or so. Why? Good question (even if it is a "why" question!). The answer seems to be related to the precipitous increase in the extraction of gas and oil from underground via use of the method commonly called fracking, short for "hydraulic fracturing." In this method, first employed in the 1940s, water mixed with other substances is pumped into the ground at high pressure to break the materials encasing the oil and gas, making extraction of these latter substances feasible. In addition to the oil and gas, considerable unwanted "dirty" water accompanies the extraction (fig. 9.6). These fracking wastewaters are then pumped back into the

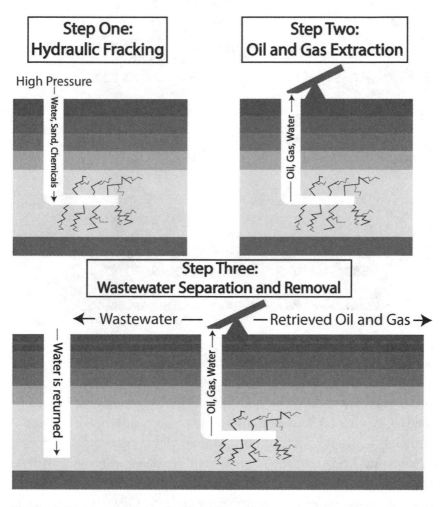

Figure 9.6. Fracking procedures. Courtesy of David Shapiro.

ground, the obvious means for disposal since the holes already exist from the extraction process.

Then come the earthquakes! As recent history shows, the increase in fracking, starting in 2009, is correlated with a dramatic increase in earthquakes (fig. 9.7). The question arises: Does fracking itself cause the earthquakes, or, say, is the pumping back of the wastewater directly responsible for the earthquakes? The answer has been controversial, with many strong

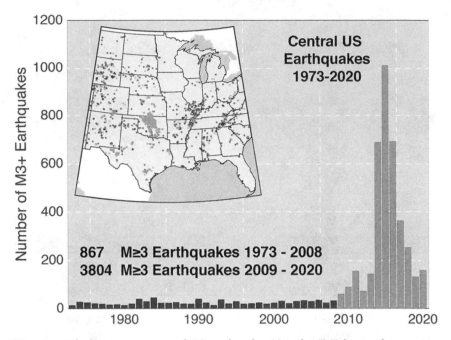

Figure 9.7. The history, 1973–2020, of U.S. earthquakes. Note that "M" denotes the magnitude of the earthquake. U.S. Geological Survey, Department of the Interior/USGS.

voices—and fewer data!—on both sides. The consensus view, backed by the U.S. Geological Survey, seems now to assign blame on the practice of pumping wastewater back into the ground. However, a study conducted in western Canada, and published in 2016 in *Science*, concluded that the cause of the earthquakes there was the fracking itself. Which side is correct? Are both correct, given the different environments? Or are the earthquakes caused by some combination? The answer is important: if fracking, then perhaps the practice should be banned; if wastewater, then perhaps another method of treating wastewater could solve the problem. Whatever, another question also needs to be asked although we do not answer it: What caused the decrease in earthquakes, starting in ca. 2015?

I do know, though, that fracking shows the direct effect human activities can have on the environment. And it presents a political issue on which citizens' voices likely will figure in our response to this problem, thus illustrating the importance of citizens being at least somewhat informed on scientific issues. Being thus informed should help them take wise positions—far, though, from a guarantee. One interesting event makes the issue more pointed than ever: a magnitude 5.6 earthquake shook Oklahoma in September 2016; no stronger quake was ever recorded in the state. Was this earthquake related to fracking, or did it have a totally independent cause?

Another serious controversy concerning fracking is the disagreement over whether it contaminates the water supply in areas near fracking sites. Here I will remark only that these problems on earthquakes and water contamination will not likely be soon resolved in a manner satisfactory to all relevant parties. Let us now tackle another of our intriguing earth studies, our knowledge of the evolution of the surface of the earth.

TEN

The Evolution of the Earth's Surface

Having grasped the history and current knowledge about the shape, size, and interior of the earth, let us turn to the history and current knowledge about the earth's surface and its possible evolution. This largely twentieth-century story is based on a collaboration, and a detective-style investigation, of specialists in geology, physics, chemistry, and biology, which yielded special fruit: a solution to the age-old problem of the evolution of the earth's surface. This illustrates by example at least one aspect of the unity of science.

In the beginning of the nineteenth century, geology was a prestigious science. Over time, however, geology lost its luster to physics, chemistry, astronomy, and later biology, and by the start of the twentieth century, the field had garnered a plodding reputation. Ernest Rutherford, a brilliant New Zealand physicist, likened it to postage-stamp collecting and described it as consisting of "making maps by identifying and locating rocks and fossils." Yet the revolution brought about by twentieth-century investigations of the earth's surface also brought with it a basic change in paradigm, reversing all of that negativity and turning it into a field of exciting potential.

Some scientists consider scientific geology to have started with the work of Nicolaus Steno, a Danish scientist, in the late 1600s. However, the heyday of geology is usually said to have begun in 1795 when the Scotsman James Hutton drew the first scientific conclusions about geology. One was that land worn by erosion is replenished by volcanism. Two more: "The present is key to the past," and "The result, therefore, of our present enquiry is that we find no vestige of a beginning—no prospect of an end." Hutton also concluded that the earth was millions of years old, which put him in conflict with the church's belief that the earth was just about six thousand years old (see chapter 12). The cornerstone of his approach was the study of rocks and strata, as was Steno's.

By the 1830s, British geologists mainly belonged to a school with members known as uniformitarians. The chief proponent of this school of thought was Charles Lyell, whose influential text promoted this view. At its base was the thought that the earth went through unending cycles, all of them about the same. In scientific circles they, the British, seemed victorious over the French, who were protagonists of the theory of catastrophes—for example, that species sometimes disappeared—thus contradicting the idea of unending, very similar cycles. This apparent victory of the British was not, however, due to the force of their logic.

Later in the nineteenth century, factions formed, each with different ideas about the surface of the earth and its changes with time. Many geologists became permanentists, with a rallying cry: "Once a continent, always a continent; once an ocean, always an ocean." An apparent example of supporting evidence was shallow seas—not oceans—that appeared and disappeared over continents, which they proposed as a means of explaining sea fossils found on land, with none found to have been a denizen of the deep sea. There were also contractionists, who believed that the earth's history was not a history of cycles but that of a one-way trip: as the interior of the earth shrank, the crust wrinkled, folded, and subsided. In this model, the motor for these changes was the earth's contraction, due to its cooling.

The underlying premises of each of these groups had problems. For example, for the contractionists, mountain ranges are far too extensive to be explained by any reasonable estimate of the effects of contraction. Moreover, radioactivity discovered very late in the nineteenth century argued strongly against contraction theory; the earth had this radioactive heat source, which would cause expansion not contraction. In a few words: the contractionists' motor had overheated.

The Idea of Continental Drift

On to the main event. In 1912, Alfred Wegener, with a doctorate in astronomy and chosen career in meteorology, took up the cudgels for an earlier proposed theory, that of continental drift. He was prompted in this proposal by his casual observation of the apparent closeness of fit between the west coast of Africa and the east coast of South America (fig. 10.1). This fit had been noticed earlier, apparently as early as 1596 after the first maps of both coastlines became available in Europe. The first publication of this match, however, is not thought to have been made until 1844.

The match was first explained in the late nineteenth century. George Darwin, one of Charles's sons, in 1878 postulated that the origin of the moon was due to an ejection of a large mass from the earth near its equator early in the earth's evolution when it was spinning very rapidly. Osmond Fisher, a geologist, four years later proposed that this mass came from the Pacific Ocean, with the Atlantic Ocean opening as Africa and South America moved apart

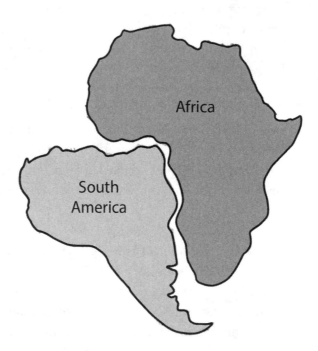

Figure 10.1. The fit between the east coast of South America and the west coast of Africa. Courtesy of David Shapiro.

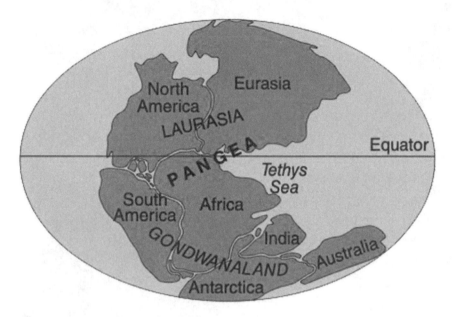

Figure 10.2. Putative Pangea, as of 250 million years ago. From Vaiden (2004). © 2004 University of Illinois Board of Trustees. Used by permission of the Illinois State Geological Survey. ("Gondwanaland" is partly repetitive; "wana" means land.)

to fill in the depression made in the Pacific. This theory of the origin of the moon is now quite thoroughly discredited and considered definitely dead. However, the subject of the moon's origin is still very much alive.

Back to Wegener. In 1915 he published a ninety-four-page book on this subject; it was entitled *The Origin of Continents and Oceans*. His theory about this origin postulated that the present continents were originally joined in a supercontinent, which he called Pangea (sometimes written Pangaea) for "all earth" (fig. 10.2). Due to unspecified stresses within the earth, he said that Pangea broke into pieces and these drifted apart, yielding the separate continents. He also envisioned later regroupings, such as the collision of India with Asia, resulting in the Himalayas.

What other evidence did Wegener adduce to support his theory of continental drift? The distribution of similar species and fossils on different continents, as illustrated in cartoonlike fashion here, would, for example, be explained naturally by drifting continents, once connected (fig. 10.3). It was hard to explain these presences of species and fossils otherwise, for

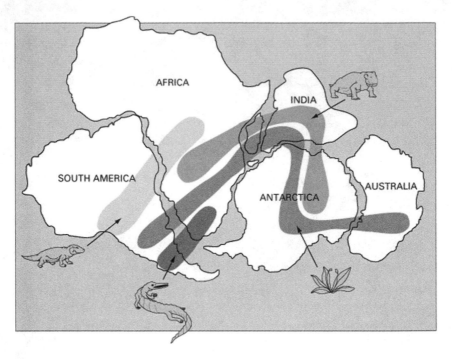

Figure 10.3. Fossil evidence consistent with drifting continents. Adapted from U.S. Geological Survey, Department of the Interior/USGS.

example, through now sunken land connections for which there was and is no evidence. Also subtropical Spitsbergen and glaciated Australia, which seemed to have occurred at the same time in the past, could then be understandable. He noted, too, similarities of mineral deposits between the east coast of South America and the west coast of Africa. Wegener also pointed to geodetic evidence that showed a westward drift of Greenland at a rate of about ten meters per year, which turned out much later to be *far* too large, by nearly three orders of magnitude (factor of a thousand)! He also noted that radioactivity would increase the temperature within the earth and would make easier the movement of continents over the earth's surface. Wegener concluded in his book on the subject, of which there were four editions, that there was only "one chance in a million that the drift idea is wrong." He developed this probability, alas not very scientifically, on the basis of the various different "coincidences."

How were Wegener's ideas received by his fellow scientists? Each and every one was attacked. Some of the best mathematical geophysicists of the day, such as Harold Jeffreys, mocked his proposal, asserting that there was no force known that could push the continents around the earth's surface as Wegener had proposed. (Jeffreys did not consider the mantle would be moving.) Many of the other critics were mostly sarcastic:

> One truly sews a rotten bag with gold thread.

> It is easy to fit the pieces of a puzzle together if you distort the shapes, but when you have done so, your success is no proof that you've placed them in their original positions. It is not even a proof that the pieces belong to the same puzzle or that all of the pieces are present.

> I found half a fossil in Newfoundland and the other half in Ireland.

Opposition to Wegener's ideas on continental drift was illustrative of opposition to new ideas that upset established views of how nature behaves. Consider the following fairly typical observation, this one in 1928 by R. T. Chamberlin: "If we are to believe Wegener's hypothesis, we must forget everything which has been learned in the last 70 years and start all over again." It is, alas, appalling that supposedly objective scientists can proclaim such illogical objections to a new idea. This opposition, by the way, was most monolithic on the west side of the Atlantic; on the European side, opinion on continental drift was far more divided.

While alive, Wegener fended off each attack, for example in his defense of inductive reasoning and his analogy with the work of Kepler and of Newton. He also tackled the alternate theory proposed by his contemporaries that a now sunken land bridge between Africa and South America allowed exchanges of flora and fauna. Wegener passed prematurely from the scene in 1930, falling victim to bad weather on a mission of rescue in Greenland in the late fall. The fourth and, alas, last edition of his book on continental drift had been published in 1929. After his death, others, sporadically, continued to battle in favor of continental drift, some with fervor. Arthur Holmes, a committed "drifter," noted that "proving Wegener wrong [on many details]

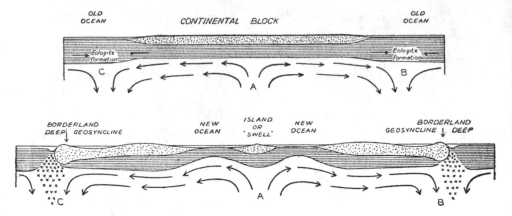

Figure 10.4. Arthur Holmes's illustration depicting his prescient idea that convection cells (indicated by arrows) of molten material in the earth's mantle comprise the motor that drives continental drift. (Ignore "Geosyncline." It is irrelevant for us.) © Henry R. Frankel 2012, published by Cambridge University Press. Reproduced with permission of the Licensor through PLSclear.

is by no means equivalent to disposing of continental drift." Wegener himself had argued that the drift was clear and that it was up to others to come up with a cause ("the motor").

Arthur Holmes not only defended drift but proposed a motor: continents are carried on the backs of convection cells, driven by the heat of radioactivity (fig. 10.4). Holmes's proposal solved a previously thorny puzzle—how the soft rock of continents could push through the hard rock of the ocean floor. It would have been like a stick of butter pushing through a steel plate, as wags of his day had proclaimed. This problem was avoided by having the continents carried along, so to say, on the backs of convection cells.

Plate Tectonics, Thermoremanent Magnetism, and the Earth's Magnetic Field

Despite Holmes's activities, Wegener's hypothesis drifted (pun intended) to the back burner of the concerns of most geologists and geophysicists in the 1930s and 1940s. While the world was largely engaged in World War II, techniques were developed for studying the ocean and for detecting small magnetic effects. Why? One main reason was to be able to detect enemy submarines. Thus, during World War II and the subsequent Cold War, more sensitive and compact magnetic detection techniques were developed, providing sensitive means to detect magnetic effects that would play a big role, scientifically, in the postwar period, especially in the earth sciences—our concern here.

One such technique is thermoremanent magnetism (TRM). What exactly is it? TRM is a property of many rocks. After these rocks arise from the liquid state through cooling and solidification of the early earth, they retain—presumably for up to billions of years—the imprint of the strength and direction of the earth's magnetic field at the time and location where they solidified. For our purposes, we can think of the earth's magnetic field as being due to a (large) bar magnet, centered at the earth's center, which changes its orientation—and strength—over time (fig. 10.5). A magnetic field is a construct of our model of electricity and magnetism (not discussed in this book). Its effects are seen, for example, in the alignment by the source of the field, say a bar magnet, of tiny pieces of iron (iron filings), as often illustrated in physics classes in high school.

Albeit in a very overly simplified manner, we can describe the main point as follows: as it cools, mainly the iron in a soon-to-be rock aligns itself with the direction of the earth's magnetic field at the rock's location; in general, the stronger this magnetic field, the greater the fraction of the iron atoms that is aligned in the rock. The larger this fraction, the greater the magnetic field produced by this iron. So the rock preserves at its formation location evidence of both the direction and the strength of the earth's magnetic field at the time the rock first solidified. This property, as we shall see, is very important to our story.

The magnetic field of the earth, remarkably, reverses its polarity (the North Pole becomes the South Pole and vice versa) on average of the order of every several hundred thousand years. This fact was discovered in 1906 by Bernard Brunhes in France and first studied systematically in the 1920s by Motonori Matuyama in Japan. The spacing of these reversals is very irregular in time, with the changeovers themselves taking place in a much shorter time, of the order of ten thousand years or less. The sun's magnetic field undergoes similar reversals, except that for the sun the time interval between reversals is now much shorter, of the order of eleven years and, at

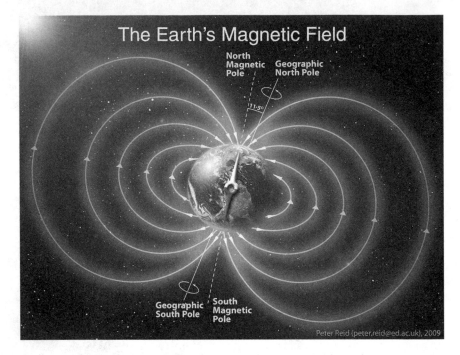

Figure 10.5. The earth's magnetic field lines. Note that, at present, the earth's magnetic axis, which points in the direction of the magnetic pole of the earth, is inclined to its geographic axis, which is the line around which the earth rotates, by about 11.5 degrees (why this particular angle at this particular time is not explained by any model of this phenomenon). Image by Peter Reid (peter.reid@ed.ac.uk), University of Edinburgh; used with permission.

least at present, much more nearly constant. Many people have heard of this sunspot cycle; far fewer know that on average its full length, to return to its same condition, is about twenty-two years, thus yielding a time scale at present for reversal about a factor of ten thousand or so times shorter than for the corresponding change for the earth.

What evidence do we have that allows us to make these claims about magnetic pole reversals? For the sun, because these reversals occur on a time scale short compared to a human life-span, we can and have made measurements from the earth of many reversals of the direction of the sun's magnetic field. For the earth, the time scale for such flips is far longer than that of humanity's knowledge of magnetic fields. What we use then are the radioactive clock ages of (nearly or actually) collocated rocks and the comparison of comparably aged rocks distributed over the globe. This project was undertaken mainly, but far from exclusively, by British scientists, such as Keith Runcorn, who devoted his professional life to this field, known as paleomagnetism (*paleo* meaning "old" or, equivalently here, "ancient").

Using TRM to Study the Earth for Fun and Profit

Magnetic pole directions determined from rocks of different ages from nearly the same location on a continent differed by large amounts. Why? Was this an indication of the wandering with time of the magnetic pole over the surface of the earth, or was it a symptom of continental drift? The same data can be interpreted equally well in either of these two ways (fig. 10.6). Of course, the interpretation could also be a combination of these two (we do not pursue this possibility further for reasons that will, or should, soon become apparent). Note that from the TRM for one (undisturbed) rock, we can infer the *direction* to the magnetic pole at the time the rock cooled but not its *distance* from that pole. We do now know though that the positions of the earth's magnetic poles, except during pole reversals, do not wander very far from those of the geographic poles, the two "ends" of the earth's spin axis.

What about rocks from different continents? Rocks of the same age from different continents yielded different pole positions that changed differently with time (fig. 10.7). Here, the obvious interpretation is that the continents moved, since otherwise the results from different continents were inconsistent, one with the other: the magnetic pole can't be at two very different places on the earth at the same time. Moreover—shades of Wegener— paleoclimatic and paleomagnetic results were in agreement, given the interpretation of continental drift.

These were heady results and had to be very robust to overcome the vast opposition. Several independent groups in different countries verified that the instruments were reliable, that the measurements were reliable, and that the interpretations were also reliable. Game over? Not on your life! But by these mid- to late 1950s, scientific opinion was finally swinging, and picking up some speed, toward drifters.

The battle then moved to the seas. After World War II, funding for ocean research also blossomed, with an accompanying explosive increase in data,

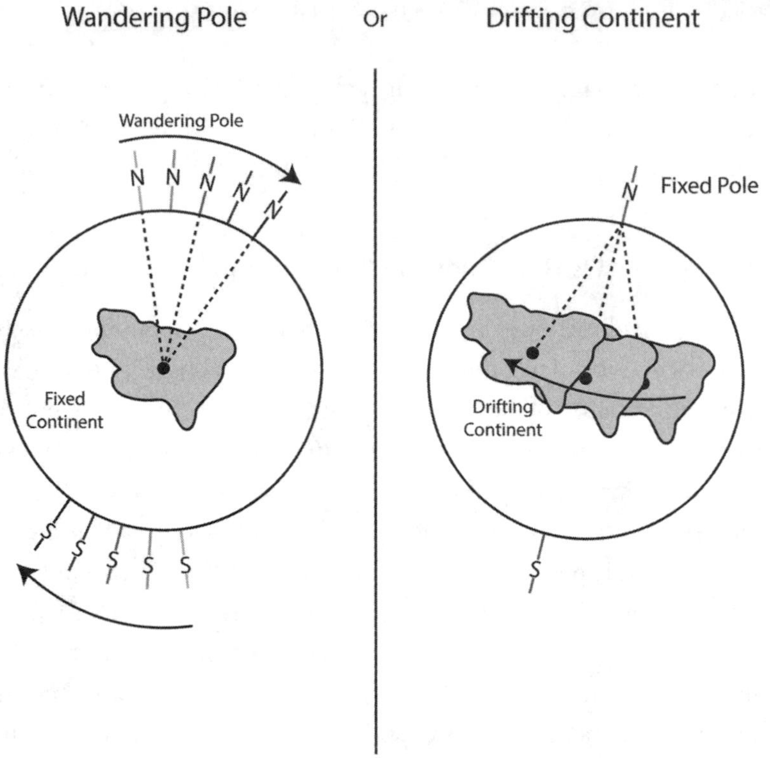

Wandering Pole   Or   Drifting Continent

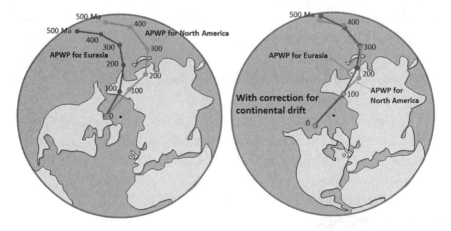

Figure 10.6. Two possible interpretations of changing direction of magnetic field: the continent is fixed on earth and the magnetic pole wanders (left), or the magnetic pole is fixed on earth and the continent drifts (right). Courtesy of David Shapiro.

Figure 10.7. On the basis of continents being fixed on the earth, the apparent position of the magnetic pole of the earth for different times (APWP, apparent polar wander path) as viewed from North America (light gray line) and from Eurasia (dark gray line) at each of these times (left panel). On the assumption of continental drift, we redraw both lines and note that they are nearly superposed (right panel). Steven Earle, *Physical Geology* (Victoria, B.C.: BCcampus Open Education, 2015). CC BY 4.0.

based on new instrumentation such as sonar (like radar, but using sound waves instead of radio waves) and on research ships that plowed through the world's oceans. The people doing this research were called, obviously enough, oceanographers. One leader was Harry Hess at Princeton. In 1960, he proposed a scenario that he called geopoetry. It was based on a major finding from mapping of the seafloor: the existence of midocean ridges, such as the one in the mid-Atlantic, which had a rift valley running down its midline (fig. 10.8). This valley was discovered by Marie Tharp, working at Columbia University during and after World War II, by making maps from data on the ocean bottom collected by many ships sounding the bottom with sonar.

As pointed out by a former student of mine, Douglas Robertson, the mere existence of this structure was powerful evidence in support of continental drift: the matching continents on opposite sides of the Atlantic could have been a coincidence; the addition of the midocean ridge elevates it to a conspiracy! Hess's proposal, or model, was that molten lava emerged from the valley between these ridges and pushed aside existing seafloor with the "ends" of the seafloor exiting back into the mantle at trenches, also disclosed from studies of the ocean floor (fig. 10.9). How did Hess's proposal play in Peoria? Not very well. The typical skeptic noted that two "wrongs" (continental drift and stable, global, mantle convection) do not equate to one "right." The old guard does not go gently into that good night.

Rhetoric aside, how could one test this idea of seafloor spreading? In 1963, Lawrence Morley (his paper on the answer was rejected by the journal to which it was submitted, because of incompetent peer reviewing, as happens occasionally) and Frederick Vine and Drummond Mathews (their paper was published) independently noted an "obvious" consequence of seafloor spreading: TRM in ocean-floor rocks should show the same "striped" pattern on *both* sides of midocean ridges, with the oldest rocks being farthest from the ridges (fig. 10.10).

The U.S. research ship *Eltanin* gathered the critical data in about 1965. This ship, with magnetometers lowered to very near the seafloor, traced out the magnetic field as the ship plowed along. The variations measured in the strength of the field on part of the nineteenth leg of the ship's research journey are shown here (fig. 10.11, top curve). (A leg of a ship's journey means,

Figure 10.8. The mid-Atlantic ridge. NOAA.

basically, a single voyage, leaving one port and subsequently arriving at the next port.) Testing of the Morley-Vine-Matthews prediction of this pattern was not even mentioned in the research proposals that the ship's voyage was undertaken to carry out. And the experimenters didn't realize, at first, what of significance they had obtained. As we can see, the pattern of magnetic anomalies is remarkably symmetric about the peak of the midocean ridge (that is, if we fold the measured curve left-right about the zero point of the horizontal axis, the two halves lie virtually on top of each other) as pre-dicted. The reversals of the polarity of the earth's magnetic field matched

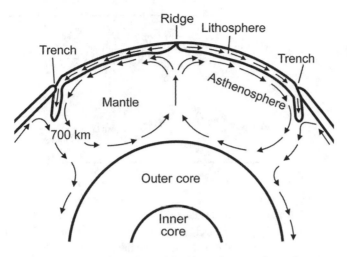

Figure 10.9. Lava emerging from the middle of a midocean ridge and creating new sea floor symmetrically on both sides of the ridge, with old sea floor returning to the mantle via trenches. Adapted from U.S. Geological Survey, Department of the Interior/USGS.

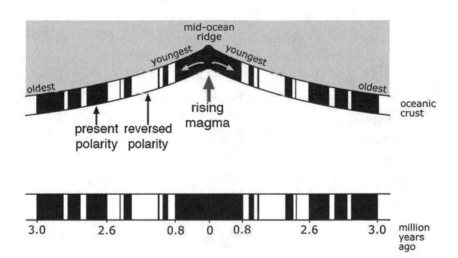

Figure 10.10. The symmetrical pattern predicted about a midocean ridge of direction of magnetic fields imprinted on rocks solidified from lava oozing from the midocean ridge. © The University of Waikato Te Whare Wānanga o Waikato. All rights reserved. www .sciencelearn.org.nz.

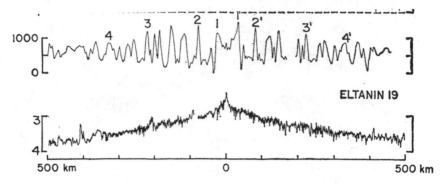

Figure 10.11. Anomalies (differences) in the strengths of the magnetic field measured in the rocks on the ocean floor on both sides of the midocean ridge in the southeast Pacific, published in 1966. The center of the ocean ridge is at zero on the horizontal scale. Note the remarkable left-right symmetry of the anomalies about this zero. (The bottom curve is the corresponding depth of the seafloor and is not of concern to us.) As for magnetic field strength: 1 gamma is about twenty thousand times weaker than the earth's current magnetic field strength of 0.2 gauss at its surface, since 1 gamma is 10^{-5} gauss. From W. C. Pitman III and J. R. Heirtzler, "Magnetic Anomalies over the Pacific Antarctic Ridge," *Science* 154 (1966): 1164–1171. Reprinted with permission from AAAS.

those obtained from surface-rock measurements, on the assumption that the lava flows from a midocean ridge were at a constant rate. These results enabled the deduction of an average spreading rate of the rocks on one side of the ridge from those on the other of about 4.5 centimeters per year averaged over more than one hundred thousand years. This deduction followed from the known time interval, which came from the radioactive dating of the comparison surface rocks, and from the extent of the measured magnetic patterns of the seafloor rocks.

Thus was the final nail placed on the antidrifters' coffin by Eltanin-19, resolving a major detective story in modern science. Continental drift, now known as plate tectonics in part because some plates involve a mixture of oceans and continents, was declared the winner by a knockout.

The View of the Earth's Surface Based on Plate Tectonics

In broad outline, the earth's surface is now considered to be composed of plates of varying sizes and shapes that move at varying rates with respect to one another, but always fitting together and, of course, occupying the whole surface (fig. 10.12). In the cartoon shown here various types of boundaries between plates are shown and described. In detail, what are some of the consequences now known about this plate tectonic revolution in our knowledge? I mention just a few to give the flavor:

1. Earthquake locations (fig. 10.13). As determined from the analysis of seismograms from the 152 seismometer locations of the Global Seismic Network, earthquake epicenters are found to be on the boundaries between plates (compare to fig. 10.12).

2. Hot spots and the origin of island arcs. An idea was put forward, and largely pushed, by Tuzo Wilson of Canada, that long-lived hot spots existed in the earth's mantle spewing up hot lava to the surface. Thus, under this theory, or model, the Hawaiian Islands were formed due to a hot spot. Since the plate that now exists over this hot spot moves roughly in the direction from southeast to northwest, the island farthest northwest (Niihau) is the oldest, about five million years old, and the island farthest southeast (Hawaii) is the youngest, only about four hundred thousand years old (fig. 10.14). These ages are from using radioactivity to date rocks on the islands.

3. Collisions between plates. As an example, the collision between India and Asia, shown here in five stages, covers the time period from about seventy million years ago to today (fig. 10.15). This collision among other consequences gave rise to the Himalayas, currently the highest mountains above sea level on the surface of the earth. Peak altitude is about 10 kilometers.

4. Current average motions of plates, worldwide. From magnetic data analogous to that from Eltanin-19 and captured worldwide, we infer the spreading rates of plates distributed around the globe (fig. 10.16). The time resolution of the data that gives rise to this figure is rather crude, of the order of one hundred thousand years or longer. Also, we do not yet have a fundamental theory that can predict reliably the values that these spreading rates have.

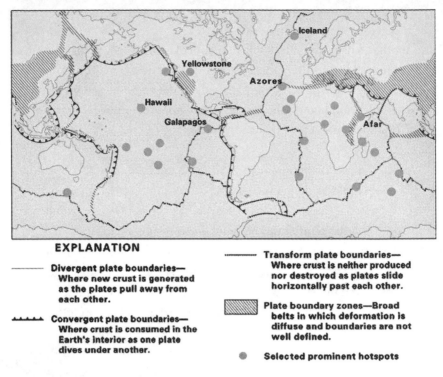

Figure 10.12. Division of the world into plates, with so-called hot spots in the mantle noted by black dots. These spots reputedly stay roughly in the same place in the earth for millions of years. U.S. Geological Survey, Department of the Interior/USGS.

World Seismicity 1976–2002

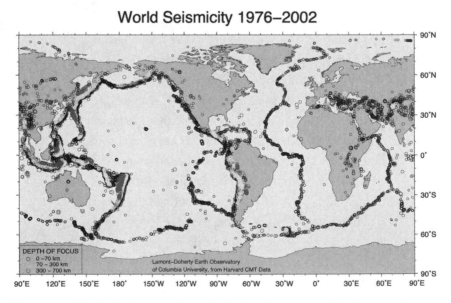

Figure 10.13. The distribution over the earth of earthquake epicenters for the twenty-six-year period 1976–2002. Won-Young Kim, Lamont-Doherty Earth Observatory of Columbia University.

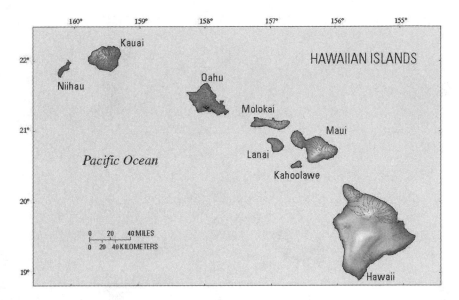

Figure 10.14. The Hawaiian Islands today. U.S. Geological Survey, Department of the Interior/USGS.

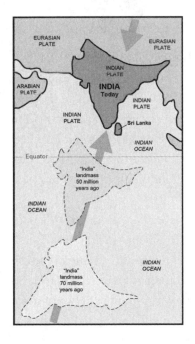

Figure 10.15 Snapshots of India on its way to and colliding with Asia, about seventy million years ago until the present time. U.S. Geological Survey, Department of the Interior/USGS.

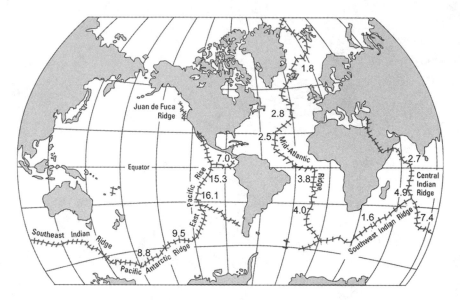

Figure 10.16. Plate motions averaged over more than one hundred thousand years, largely from magnetic data. The numbers have units of centimeters per year. U.S. Geological Survey, Department of the Interior/USGS.

Measuring Contemporary Plate Motions

How may we obtain contemporary, fine-time-resolution plate motions? Technology developments over a half-century ago paved the way to an answer. There was a competition to measure contemporary plate motions between two techniques, and between corresponding protagonists. One technique goes by the ultra-sleek name of very long baseline (radio) interferometry (VLBI), and the other is called satellite laser ranging (SLR). The idea to use the second technique followed that for the first (which first was, in fact, mine in 1966).

How does VLBI work? The basic idea is to observe with radio telescopes very distant celestial objects such as quasars, which emit radio waves. Being so distant, these objects have no detectable motions in the plane of the sky and so appear motionless; they are in effect stationary reference points. Simultaneously observing such a celestial object with antennas at two or more locations on the earth, we detect at each antenna the same radio waves as at the other antenna(s), with a key difference: we receive the same waves, but at different times, depending on the difference in distance of the two antennas of each pair from the distant source of the waves. (Keep in mind that, relative to the distance between us and the radio source, the distance between the antennas on earth is truly minuscule.) But, you might well ask, how can we know the differences in the times of arrival of the waves at two well-separated antennas without having at each of the two locations exceedingly good clocks? We can't. The clocks used are highly accurate and their variations are modeled in rather sophisticated manners. To determine the distances between each pair of antennas, as well as the directions to the quasars and the clock behaviors from these measurements of time differences, requires a fairly intricate calculation and we therefore skip a presentation of the details.

The current overall accuracy achievable is now so high, for example, that with a large array one can determine the separation of two antennas on different continents with an uncertainty of the order of a few millimeters from several hours of observations of several different sources, a relative accuracy of about one part in a billion!

The principles of SLR are far easier to explain than are those for VLBI. One determines the distance to a satellite by transmitting a short pulse of light from a laser toward the satellite and detecting the reflection of that pulse from the satellite back at the telescope used to transmit the short pulse. The satellite is equipped with what are called corner reflectors (or corner cubes), which reflect light back in the direction from whence it came. There are many satellites currently in orbit that are festooned with such corner cubes and hence are targets for SLR.

Whereas the sources of radio waves used in VLBI appear stationary on the sky, the targets of SLR are very much nonstationary. This fact is both a pain and a panacea for SLR: from measurements of the round-trip times of the laser signals to the various satellite targets, one must determine not only the locations of the SLR sites but also the orbits of the satellites and— the panacea part—the gravitational field of the earth, which makes itself known through the perturbations (small changes) it induces in the satellite orbits.

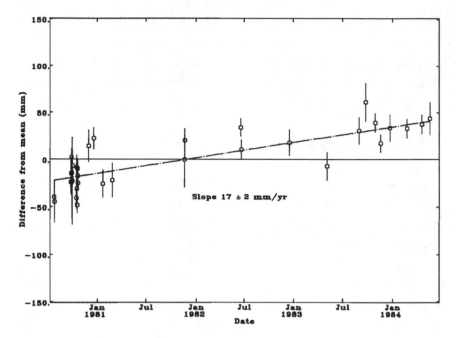

Figure 10.17. The first determination of contemporary plate motions, done via VLBI observations between the Haystack Observatory in Massachusetts and the Onsala Space Observatory in Sweden. This early result has stood the test of time. Courtesy of Thomas Herring.

Which technique—SLR or VLBI—was the first to yield reliable results
for measurements of contemporary plate motions? Leaving aside early false
claims (see below), the winner by several years was . . . VLBI, in 1985 (it
took almost nineteen years from the date of the idea to that of obtaining a
reliable result!). The result was a determination of the change in the distance between the Haystack Observatory in Massachusetts and the Onsala
Space Observatory in Sweden of 1.7±0.2 centimeters per year, a value that
matched, to within the uncertainties, the long-term average based on the
measurements of sea-floor spreading (fig. 10.17).

Before leaving this subject, I will add some comments of sociological
interest. Preceding our successful determination of contemporary plate motions, there were false claims of detections of such motions made by other
VLBI groups and by the SLR group. I will mention in detail only one, especially egregious, episode. There was a group attempting to measure, via

VLBI, plate motions across the Pacific. This group presented results from several years of observations that, figuratively, knocked my eyes out. They showed measurements of distances between sites versus time, with all the points fitting remarkably close to a straight line, but with error bars on each point that were huge compared to the distance of the point itself from the straight line. I calculated quickly that were these points reliable, obtaining such results from random errors would occur about once in every 10^{18}th time—once in every billion billion!—a likelihood so small as to be totally unlikely to have occurred. I was puzzled, because if this group were cheating, they could easily have done a far better job that would not have immediately raised such suspicions. Never the mind, I expressed my concerns to the head of that group and got nowhere. So, I told him that I would speak to his supervisors, which I did. I noted first that, on the surface, I had a conflict of interest as I had initiated this race to measure contemporary motions and was very much involved, but I went on to urge them nonetheless to listen to my arguments and then to themselves look into the issues that I raised. They did nothing. I persisted. After a *year* of pestering, I finally succeeded in getting their original data for our group to examine. These data showed the complete absence of useful information. All of their results were simply made up. The supervisors then fired the head of the team. I believe it took this now former group head less than two weeks to find another position at twice the salary. End of story. This type of behavior is very rare in science but, alas, not without precedent.

Now for the last question—of this series only! How do we make observations of plate motions today, more than thirty years later? Like (almost!) everything else these days, we make use of the GPS (Global Positioning System) satellites, as a former student of mine, Charles Counselman, and I first suggested, independently, in the late 1970s; Chuck developed the first receivers for this purpose, no mean feat. The GPS satellites transmit radio signals that are very strong as received on the earth compared to those from the exceedingly distant quasars. Thus, instead of the huge antennas needed in VLBI observations of such cosmic sources of radio radiation, only very tiny—and correspondingly inexpensive—antennas are needed to detect on the earth the radio signals from GPS satellites. The cost of each GPS ground terminal was originally under about \$10,000 instead of being about \$10 million at least for a VLBI ground terminal; it is now near \$2,000. Because of this dramatic reduction in cost, the world is now being covered with thousands of GPS ground terminals for geodetic studies, which use the same principles as for the VLBI technique. With GPS one must also estimate the orbits of the satellites from the data, which is onerous but with modern computers is handled routinely. One might say, using the vernacular, that this development and use of GPS interferometric ground terminals has become a worldwide cottage industry. These ground terminals are not disported uniformly; their greatest concentrations are near earthquake-prone zones in high population regions, such as near the San Andreas fault in Southern California and in various parts of Japan.

Are VLBI systems now obsolete for earth-related studies? No; they are used primarily to study (small) variations in the rotation of the earth. Such study requires a fixed reference frame as is provided by distant, compact cosmic radio sources such as quasars but not by GPS satellites. In some places, such as Hartebeesthoek in South Africa, there are collocated a VLBI system along with an SLR and a GPS system.

Having given an uneven (in terms of pages devoted) treatment of the shape, size, structure, and evolution of the earth and its surface, we move on to discuss its age—perhaps a welcome change from the intricacies of VLBI and SLR. But this discussion will encompass a significant number of its own intricacies. In fact, as a precursor to telling the story of our solving the problem of the age of the earth, the next chapter describes the uncovering of some of the science that was used to obtain that solution.

The incredibly rich paradigm of plate tectonics has invigorated earth scientists and revolutionized their science for the past more than five decades and has not yet disclosed any signs of slowing up, although slow up it must, and will, given sufficient time.

ELEVEN

The Structure of Matter

This chapter may seem an odd one to place here and deserves explanation. In addition to providing the necessary scientific background to understanding our long investigation into the age of the earth (treated in chapter 12), this chapter tells a fascinating story: I will describe details about discoveries that gave us a deeper level of understanding of the structure of matter, useful for understanding some parts of our discussion of the age of the earth.

In the last quarter of the nineteenth century, back in the physics laboratory, amazing properties of nature were being unfolded, of course building on earlier work. It was known since the late eighteenth century from investigations in chemistry that there were different kinds of matter, called elements. For example, the great French chemist Antoine Lavoisier, a guillotine victim of the French Revolution, discovered some of the properties of oxygen. Each of these elements came in the form of identical particles called atoms. What were these atoms, and how did we know they existed?

The person usually credited with being the first to postulate the existence of atoms as the smallest particles of matter, not further divisible, was the Greek philosopher Democritus (ca. 480–390 BCE), based not on experiment or observation but on pure reason. John Dalton (1766–1844), an English scientist, is usually credited as the first to establish the existence of atoms from inferences based on the observed behavior of matter. Even given our acceptance of atoms as the smallest particles of matter, what does

that tell us of their structure? Nothing! Let's look first at a few simple observations that we can make that imply atoms, if they exist, are particles with
spaces in between them.

Liquids, such as water, seem incompressible. We can push down as hard
as we possibly can on a liquid surface—when the liquid is in a very strong
container!—and the surface does not give; there is no discernible compres-
sion. One could be excused from concluding that, for example, water is
solid, through and through. Yet if one pours into the water a colored dye, it
spreads quickly, and evenly, throughout the water. We find the same result
by pouring sugar into the water and stirring it: no matter which part of this
water we taste, it tastes sweet, as sugar water should! Similarly for salt and
many other powdered substances. How are these results possible? The water
must be penetrable. It seems likely to be composed of invisibly small parts
with spaces in between them.

Another indication that liquids are made up of smaller parts was discovered in 1827 by Robert Brown—so-called Brownian motion. Brown noted
that tiny particles in water jiggled around by small amounts, discernible
when he looked at the particles through his microscope. What is pushing
these particles around? The obvious answer is the smaller parts of which the
water is composed. Einstein in 1905 was the first person to mathematically
analyze the motions of these particles in terms of small particles of which
the water is likely composed. His mathematical model correctly predicted
the statistics observed for the motions of the tiny particles that were placed
in the water.

But what of the structure of the atoms—presumably composing the small
particles that make up the water? (We speak of atoms. But, for example, in
water as we now know, the atoms are combined into entities, which we call
molecules. Each molecule of water is composed of two atoms of the element hydrogen and one of the element oxygen; this combination is tightly
bound, and each exemplar is considered as one entity, a molecule.)

Back to our questions: How, for example, can we account for the unique
signatures atoms (somehow) impart to the spectral lines through the specific frequencies of the lines the atoms emit? We have here a remarkable
detective story in the making. As is often the case, technology spurred the

solution. At that time, near the end of the nineteenth century, it was apparently a common belief among physicists that everything fundamental about the natural world had been discovered—only some cleaning up was needed—despite the existence of basic effects about whose understanding not a soul had any inkling. Then new tools were developed with which the basic properties of matter were explored in physics laboratories. These new tools, in the hands of clever people, soon resoundingly defeated this notion that nothing new and important was left to discover about nature's behavior at a fundamental level, as we now describe.

A key new tool for carrying out such fundamental physics experiments was the Crookes tube, a forerunner of the cathode-ray tube used in early televisions and named after its inventor. The tube's main characteristics were the high voltage that could be created between the metal plates, one at each end, called the cathode and the anode (see below), the high vacuum inside of the tube about one ten-millionth of the normal air pressure outside the tube, and, finally, the transparency of the (glass) walls of the tube.

This Crookes tube was used by the English scientist J. J. Thomson in experiments. He applied a high voltage between the cathode and the anode, which produced what were then called beta rays that traveled from the cathode to the anode. The paths followed by these beta rays were affected by magnets. For example, when Thomson placed the pole, say, of a bar magnet near the Crookes tube, the path followed by the beta rays was deflected in a certain direction, say downward, and when, instead, the other pole of the magnet was pointed in that same direction, the path of the rays was deflected in the opposite direction, upward. These deflections were seen because of the light these rays created in interactions with the residual gas in the Crookes tube. From knowledge of the energy of these rays and their angle of deflection caused by magnets of known strength, Thomson was able to calculate that the particles composing beta rays each had a mass about one two-thousandth the mass of a hydrogen atom. Thus was the electron discovered, the first of the so-called elementary particles, in 1897. Lest we lose sight, we are in the midst of describing the discovery of the structure of the atom, which plays a key role in much of the later parts of this book.

Thomson devised his "plum pudding"—interestingly, in American speak, the English plum is called a raisin—model of an atom, based on his discovery of the electron, which has a negative charge. (Two signs of charge are known; they are called positive and negative. Benjamin Franklin introduced the "positive" and "negative" terminology for charge; based on his choice, the electron has a negative charge.) Thomson's model had a uniform distribution of positive charge (the pudding), with an equal total charge as the electrons (the plums or raisins), which were distributed—somehow!—throughout the atom. A main point: this model of Thomson's had *no* explanatory power, other than to explain why atoms seemed to have no net charge. For example, it was incapable of explaining the spectra of atoms, even of the hydrogen atom, which being the lightest was presumably the simplest of all atoms. It also was not clear that this model was stable; it did not address the issue that opposite charges attract, and so it gave no description of how the electrons and the positive charges remained separated. We were still in the dark as to the nature—structure—of the atom. Thomson is nonetheless revered as a scientist, mainly because of his fundamental, key discovery that beta rays were extremely light particles, each with the same negative charge.

No important progress was made in unraveling the mystery of the structure of the atom for about another dozen years. Then, in 1909, a new type of experiment was used with a startling result. The lead experimenter was the physicist Ernest Rutherford, who had come from New Zealand to Cambridge, England, to practice his profession, since Cambridge was at the center of new discoveries about atoms. The new type of experiment was called a scattering experiment. This name signified that a beam of particles was directed toward a target; the beam interacted with the target and left the interaction region with the individual particles comprising the beam going in different directions from that in which they started—that is, they were scattered. By analyzing the directions in which these particles were scattered, one could infer basic properties of the target. Over time, scattering experiments have used beams of particles traveling at higher and higher speeds. The higher the energy of collision, which increases with the relative speeds of the particles, the more deeply the targets can be probed. For example, if a beam of particles moving very slowly were to hit you, the particles might be stopped by your skin and not probe anything about your innards. If, on the other hand, the beam of particles traveled fast enough—had enough energy—to penetrate your skin, the particles could investigate properties of your insides, disclosed by how they scattered from your insides. Similarly, in scattering experiments, physicists continue to probe the basic properties of matter at higher and higher collision energies. The latest major triumph in this game was the discovery of the Higgs boson, mentioned in the Introduction.

Back to Rutherford: to him can be attributed the start of this long parade of ever more penetrating scattering experiments. His pathbreaking experiment consisted of directing a fast-moving beam of so-called alpha particles head-on into a thin foil of gold (fig. 11.1). Each of these positively charged particles—each with twice the charge of an electron, but of opposite sign—is identical with the main part of a helium atom. The particles for this experiment were obtained from the radioactive decay of the element radium (discussed in the next chapter). This experiment was carried out by Rutherford's group, in particular by Hans Geiger and Ernest Marsden. They had arranged suitable detectors, which had the capability to record alpha particles scattered into an evacuated region surrounding the source of the particles and the gold foil.

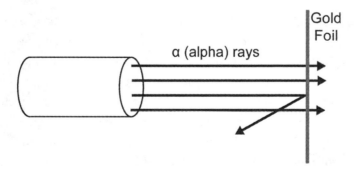

Figure 11.1. Scattering of alpha particles by thin gold foil, arranged perpendicular to the direction of the incoming particles. Courtesy of David Shapiro.

The results from this scattering experiment were totally unexpected. Almost all of the alpha particles traveled through the target virtually undeflected. A very small fraction was scattered nearly backward toward the source of the alpha particles. What could this very unexpected result possibly mean? Rutherford, after some thought, had the answer: the atom must be mostly empty space, but with a very hard, small, massive core or, as we now refer to it, a nucleus. Given that the atom was overall neutral in charge and had so many electrons, it was reasonable to conclude that the nucleus had a positive charge that just counterbalanced the charge of the electrons in the atom.

Published in 1911, a little over a century ago, the paper reporting this result of Rutherford and his team marked a major advance in our knowledge of the structure of the atom. To quote Rutherford:

> It was quite the most incredible event that has ever happened to me in my life. It was almost as incredible as if you fired a 15-inch [artillery] shell at a piece of tissue paper and it came back and hit you. On consideration, I realized that this scattering backward must be the result of a single collision, and when I made calculations I saw that it was impossible to get anything of that order of magnitude unless you took a system in which the greater part of the mass of the atom was concentrated in a minute nucleus. It was then that I had the idea of an atom with a minute massive center, carrying a charge.

Clearly, Thomson's plum (or raisin) pudding model of the atom needed to be replaced. We now had a model of an atom with a positively charged, exceedingly small, massive nucleus, with negatively charged, light electrons making up the rest of the atom. How did it work? According to classical physics, in which charges with opposite sign attract each other, electrons in fixed starting positions inside such an atom would very, very quickly accelerate (fall toward) and collide with the nucleus; the atom would then presumably go something like "poof." Thus, this structure of the atom would be unstable and couldn't exist, as we would conclude from our knowledge of the model of physics that seems to work well on a macroscopic level—

very similar to a main criticism leveled against Thomson's plum pudding model. Ah, but suppose the atom were like a miniature solar system, with the electrons in orbit about the central nucleus. Such a model would also be unstable: were it correct and were classical physics of that era correct, the atom would collapse in an exceedingly short time. This collapse is opposed to the situation in the solar system, which is stable on time scales far, far longer than that for which the human species is likely to exist. Our solar system is predicted to take so long to collapse because gravitational radiation by planets is expected to be so incredibly weak compared to the light radiation predicted to be shed by an electron moving in orbit about a central nucleus. (For an inkling of a comparison of their relative strengths, see the relevant discussion near the end of chapter 7.) Why would such an atom be expected to collapse so quickly? Because a charged particle like an electron radiates light if it is in orbit—that is, accelerating (through constant changes in its direction of motion). Thus, the electron would very quickly lose its orbital energy through radiation and crash into the nucleus. The atom would again go "poof." Something strange was clearly going on. Nature's behavior on this atomic level must be very different from what we had thought, based on our observations of nature's behavior on a macroscopic level. What was needed was a new idea.

The Modern Model of the Atom

Enter Niels Bohr (1885–1962), then a newly minted Ph.D. from Denmark. He was also attracted to Cambridge as the center of new physics. Bohr proposed a profound new twist to the idea of a miniature solar system model of the atom. He proposed drastically different properties for this system, based somewhat on a few related ideas floating around at that time. The first was that electrons can move only in certain specific orbits, with the nucleus at the center, and that they can (somehow!) jump from one such orbit to a neighboring orbit, either by absorbing or emitting a photon. (What is a photon? As mentioned earlier, our model for light is very complicated; sometimes it behaves as if it were composed of waves and sometimes as if it were composed of particles. When it behaves like particles, light is said to

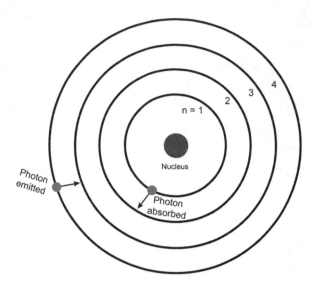

Figure 11.2. Sketch of Niels Bohr's model of the atom (the four orbits shown are not at their proper relative distances from the nucleus for this model). The gray circles represent electrons orbiting about the nucleus, with the arrow pointing to their new orbit after emitting or absorbing a photon. Courtesy of David Shapiro.

be composed of photons, with each photon particle corresponding to a particular frequency and having a particular energy. Seemingly weird to those raised on macroscopic views of nature, this model was nonetheless viable to represent some of nature's behavior on a truly microscopic level.) The second property was that a jump to a lower-energy orbit is accompanied by the emission of a photon to conserve energy, and similarly the absorption of a photon is accompanied by a jump to a higher-energy orbit. Both processes are illustrated here, with the allowed orbits of the electrons labeled numerically, and the lowest-energy orbit ($n = 1$) being closest to the nucleus (fig. 11.2). These special orbits are chosen to satisfy certain conditions that we will not here explore. One goal was that the setup should yield agreement with the observed spectrum accompanying radiation—photons— emitted from, and absorbed by, each type of atom.

These postulated properties for the Bohr model of the atom represented quite a departure from previous thinking. All well and good, but did Bohr's model achieve its goal of explaining atomic spectra? Recall that the previous model postulated by Thomson had no explanatory power whatsoever in regard to spectra. Bohr's model yielded the observed spectrum of lines observed for emissions (and absorptions) of hydrogen atoms, reproducing theoretically the empirical model for this spectrum that had been developed a generation earlier by Johann Balmer and generalized three years later by Johannes Rydberg. But Bohr's model failed to accurately predict spectra of atoms of elements heavier than hydrogen. A still better model was needed.

A dozen years later, this better model made its appearance. In 1925, two physicists, Werner Heisenberg and Erwin Schrödinger, separately developed a vastly more complicated model, each in a different mathematical guise; later these two formulations were shown to be equivalent. This model is called quantum mechanics and takes over some of the features of Bohr's model. It has proven amazingly accurate in its predictions of all atomic and molecular phenomena, even when tested at levels of one part per trillion! I believe that, in this sense, it, with the general theory of relativity, are two of the very few most accurate models of nature so far developed by humanity.

Did we then, after the invention, or discovery, of quantum mechanics, have a good picture of atomic structure and its behavior with time? Yes, but puzzles remained. Although we did not yet mention it, the mass of an atom is usually greater than twice the mass of the number of positive charges in the atom, for sufficiently massive atoms, assuming that each positive charge is that of the nucleus of a hydrogen atom. Also, some atoms, which have seemingly identical chemical properties, have somewhat different masses. How were these smaller mysteries cleared up?

The Discovery of the Proton, of Isotopes, and of the Neutron

The discovery of the proton is not so easy to attribute to a single individual. After the discovery of the electron, scientists realized that positively charged entities must exist. Why? Simple: macroscopic matter is neutral in charge (that is, it has no net charge). Therefore, something must exist to counteract the negative charge of the electron. Only about twenty years later did this entity become clear. Continuing with scattering experiments, Rutherford directed energetic—fast-moving—alpha particles into nitrogen gas and detected hydrogen nuclei coming out. This result led to the hypothesis, supported by later experiments, that the hydrogen nucleus was a single, positively charged particle, which was named the proton. Further, the proton was a constituent of all atomic nuclei—that is, of all of the different elements. Each element was given a number, its so-called atomic number. This number is equal to the number of protons in the atom's nucleus. Thus, hydrogen has the number 1 and uranium the number 92; uranium is the heaviest—most massive—of the elements that occur naturally in the earth. Some heavier elements have been created by humans in high-energy scattering experiments; all of these latter are very unstable and have extremely short lifetimes.

There were more problems. The masses of many different elements, for example, were not close to integer multiples of the hydrogen atom. How could that be if the nuclei were composed of protons and if, as first shown by Thomson, the electron's mass was so tiny compared to that of the hydrogen atom? Further, as noted above, these masses, especially for elements with numbers above about 20, exceeded the total for the protons in the nucleus by more than twice the total mass of the protons—that is, the nucleus was more than twice as massive as were the protons in the nucleus. What could be going on?

Frederick Soddy was the one who discerned that some atoms of an element showed differences in mass from most other atoms of that element, but could not be distinguished chemically; all of these atoms seemed to behave in chemical reactions exactly the same way. They were termed "isotopes" of one another. He noted, "Put colloquially, their atoms [that is, those of isotopes] have identical outsides, but different insides." What he was saying, in effect, was that there was some other constituent that was electrically neutral and did not noticeably affect the atom's chemical properties, which were apparently affected only by the charged particles in an atom. What could this neutral entity be? Extra electrons and protons combined, with equal numbers of each, were soon ruled out by various (arcane for this book) arguments.

Not until 1932 was it demonstrated by James Chadwick, again via scattering experiments, that a neutral particle—one with no electrical charge—

could be knocked out of a nucleus of an atom. He was able to estimate the mass of this particle, obtaining a result within about one part in two thousand of the modern value. The mass he obtained was approximately 0.1 percent heavier than the proton. Since it was electrically neutral, this particle was named the neutron. So, the problem of the discrepancy between mass and charge of atomic nuclei was solved. But why particular atoms had the number of neutrons in their nuclei that they had was—and mostly still is—not known: there is no good general model.

Interestingly, whereas the neutron is (mostly) stable inside the nucleus, when knocked from the nucleus it is very unstable, decaying in about ten minutes into a proton and an electron. The proton, by contrast, does not seem to decay and has been determined to be stable for at least 10^{34} years. Since the universe itself seems to be much less than 10^{11} years old, how could we possibly tell that a proton's lifetime is at least 10^{34} years? It's because we deal with so many protons in these experiments to determine a proton's lifetime that we can infer from not a single one having decayed over the time of the experiment that the average lifetime is at least as long as the limit given above.

At last we have a model for the structure of an atom that agrees with our observations! In this model, the atom has a very compact, massive, and positively charged nucleus, composed of protons and neutrons, with the former related to the chemical properties of the atom; electrons, in equal number to the protons, surround the nucleus and control the chemical properties of the atom, leaving the atom overall neutral in its charge. Different elements are different in their chemical properties because they have different numbers of electrons (and, of course, of protons).

The Discovery of Radioactivity

We now retreat to a time nearly a generation before Rutherford's determination of the structure of the atom and a fortiori before Bohr's model of the atom. In 1896, Henri Becquerel, following on the heels of the discovery of X-rays, studied materials that emitted radiation after being exposed to sources of energy, such as sunlight. He had been studying a uranium salt this way. One rather cloudy day, he placed the salt in a drawer where he had also stored an unexposed photographic plate. For some reason, he later looked at the plate and found that it had darkened. What could have been the cause? He studied the emissions of the uranium salt, discovering that they darkened a photographic plate, even though not exposed to sunlight. Further, he then employed magnets, which deflected the emissions, establishing that they were charged particles. From the direction of the deflection, he could tell the sign of the charge—positive or negative.

Becquerel was joined in his studies by his student Marie Curie and her husband, Pierre. In her investigations, Marie discovered the element radium—which she named. It had properties similar to that of the uranium salt. She coined the term "radioactivity" for this type of phenomenon. It stuck.

Let's step back now and look at what happened from our modern perspective: elements, which previously had been thought immutable, were here being transformed to other elements through the emission of particles, presumably from their nuclei. (Remember that the nuclei of atoms were not discovered until about thirteen years after the discovery of radioactivity.) Only very heavy elements, such as radium and uranium, seemed to have this property; we now know that some isotopes of light elements also exhibit radioactivity. How did this so-called radioactivity of certain elements fit into our model of the atom at that time? It didn't. Absent any understanding, scientists set out to determine the properties of radioactivity. They found strange behaviors. Taking a large collection of atoms of a specific radioactive element (it was hard not to have a huge collection of atoms: they are extremely small compared with the smallest [microscopic] amount then possible to separate), they discovered that a certain fraction of them decayed into another element after a certain time interval. For example, they found that each radioactive element had a half-life. That is, after the passage of a specific amount of time, unique for each element, only half of the collection of atoms would remain in their original condition. We call this amount of time the half-life of the element. Which specific atom will decay at any given time? We do not know; we know only statistically the overall number that will decay in a specific amount of time for a large collection of atoms

of a given radioactive element. Very strange, but that is the way nature behaves. Into what does the element decay? Each radioactive element decays
into another element, which itself may or may not be radioactive. If it, too,
is radioactive, it will decay with its own half-life into yet another element.
In fact, the chain of successive new radioactive decay products can be quite
long, ending only when a stable element is the decay product.

Matters are even more complicated. In some elements, the decay can be
into either of two different elements, with the "choice" again being statistical. For example, a radioactive element can be twice as likely to give rise
upon decay to the element A than to the element B. In such a case two-
thirds of the decay products will be element A and one-third element B.

When a radioactive element decays, what comes out? Good question.
My weasel-worded answer is: different things for decays of different radioactive elements. To be more enlightening, I point out that for some of the
decays, an alpha particle is emitted and for others a beta particle. As we
already know from J. J. Thomson's work, beta particles are what we now
call electrons. Alpha particles, on the other hand, were found to be far, far
heavier (approximately eight thousand times heavier!) and to be essentially
identical to the nuclei of helium atoms, as also noted above.

Radioactivity is an alchemist's dream, the conversion of one element (specific type of atom) into another element; the alchemist might have dreamed
of changing iron into gold on demand, but nature, via radioactivity, does
not work that way, although it does change some elements into others. This
naturally occurring changeover, or conversion, of an element into another
element is not a property of all atoms but only of some, mostly the more
massive ones. The actual changeover occurs randomly among atoms of a
given type, but on an overall reliably predictable schedule, because of the
truly huge number of atoms involved in any macroscopic quantity. Thus,
a gram (only 1/453rd of a pound) of uranium contains over 10^{20} individual
atoms—more than a hundred billion billion!

Because of the large numbers of individual atoms of any given radioactive element in an "ordinary" sample, we can determine in the laboratory in
a relatively short time a very accurate value for its half-life, no matter if that
half-life is extremely long as some are—up to several tens or so billion years.

Table 11.1. Four pairs of radioactive parent and stable daughter elements

Potassium 40 (19)—Argon 40 (18):	1.25 Gigayears (billions of years)
Rubidium 87 (37)—Strontium 87 (38):	48.8
Thorium 232 (90)—Lead 208 (82):	14.0
Uranium 238 (92)—Lead 206 (82):	4.5

Note: The number following the name of the element is the number of neutrons + protons in the nucleus; the second number (in parentheses) is the atomic number of nucleus—that is, the number of protons in nucleus. Decay chains are of different lengths and complexities.

To demonstrate, let's look at four pairs of elements (table 11.1). The first element of each pair is the radioactive one, the second element being the final stable product. The length of time given after each pair represents the half-life for the decay of the first element into the final one of the chain, the stable end product.

In this process of radioactive decay, the particle emitted from the atomic nucleus at each step has energy, due to its motion, and hence imparts heat to its surroundings—a source of heat not previously recognized but exceedingly important for estimating the age of the earth based on cooling arguments. Different radioactive elements in general give off greatly different amounts of heat/energy in their decay. Heat, by the way, is a form of energy; heat is due to the motions of the constituents of the material to whose heat content we are referring.

We have now prepared the background to treat the problem of the next chapter: What is the age of the earth?

TWELVE

The Age of the Earth

"There are few problems more fascinating than those that are bound up with the bold question: How old is the earth? With insatiable curiosity men [*sic!*] have been trying for thousands of years to penetrate the carefully guarded secret," said Arthur Holmes in 1927 (after the solution was in hand).

Before the twentieth century, no one had a clue. The age of the earth was truly a vexing problem; obvious handles were missing. Ancient Greeks, to my knowledge, left no evidence about their views on this point, other than the belief that the earth always was, a belief supposedly held by Aristotle. The biblical story of Genesis provides an ancient scenario for the creation of the earth. How, however, can one turn this biblical story into a corresponding numerical answer for the age of the earth? Out of many attempts, one became more famous than the others, apparently because it was entered in the margin of a copy of the King James Version of the Bible and was, somehow, subsequently entered in all—or, at least, many—annotated versions of that Bible. Thus, its fame spread. This attempt was made by Bishop James Ussher in 1650, who analyzed the Bible's "begats," among a considerable amount of other data, and concluded that the earth was created in the evening on October 23, 4004 BCE (Julian calendar).

The first scientific approach to determining the age of the earth is usually credited to Benoît de Maillet (1656–1738). He thought that he could use changes in sea level over a period of seventy-five years, measured by his grandfather, plus his own observations of the earth from his extensive travels, to estimate the earth's age. The result was a number of different estimates based on various lines of argument, with the oldest age being about two billion years. De Maillet opposed the clerical argument for a young earth. But he, like many of his contemporaries and predecessors, had a great fear of the church's reaction to his estimate. He thus wrote up his results as a conversation between a French missionary and an Indian philosopher, Telliamed—his name spelled backwards—and his conclusions about the age of the earth were not published until he had been dead for a decade.

Next up in the scientific realm was Edmond Halley, still of comet fame, early in the eighteenth century. He had a clever idea, based on his question: How long would it have taken for the oceans, starting with freshwater, to reach their present level of salinity? How to approach obtaining the answer? Halley proposed to consider all rivers that flow into the seas and to measure the rate at which they carry salt into the seas. No doubt a clever idea, but also no doubt a flawed one. As one prime flaw: What is the basis for saying, or even thinking, that the rate at which salt now flows into the seas has remained constant over the age of the earth?

We consider now Isaac Newton and his contemporary, and mathematical rival, Gottfried Leibniz. Both, apparently independently, reasoned that the age of the earth could be estimated through calculation of the time it took the earth to cool from an initially high temperature to its present surface value. Newton and Leibniz seemed to be aware that this cooling time, t_{cool}, for a body of radius, R, obeyed:

$$t_{cool} \propto volume/surface\ area \propto R^3/R^2 \propto R, \qquad (12.1)$$

since the heat energy that the earth contained should be proportional ($\propto$) to its volume, which varies with the cube of its radius, R, whereas the heat it could lose, or get rid of, would of necessity leave through its surface, whose area varies with the square of its radius. The division of the first by the second is proportional to the time for the body to cool. In other words, the amount of heat it contains divided by the rate at which it leaves yields the time it takes the body to cool. The result of the division shown in equation 12.1 yields a value proportional to the time a body takes to cool. The actual time it takes to cool depends on the values of other quantities, too, such as the actual rate of heat loss. According to this result, though, the larger the object, the longer it would take to cool down from a given temperature, all other characteristics being the same. As far as I know, neither Newton nor Leibniz carried this idea any further. In fact, the necessary model of heat transfer to carry out such a calculation probably did not then exist.

The next step was experimental. Comte de Buffon (1707–1788) decided to measure the cooling times of ten white-hot iron spheres and of various other spheres made of supposedly earthlike materials. He would then extrapolate to the earth's size from the results for his spheres, which ranged in radius from approximately 1 to 15 centimeters in equal steps. He set up his spheres to cool in his basement. His results indicated a linear relation between cooling time and sphere radius, as one could infer from equation 12.1. He then scaled up his results to imitate the earth's cooling from an assumed initial melting temperature of iron. His result for the age of the earth was approximately seventy-five thousand years, more than tenfold higher than Bishop Ussher's value. Despite his experimental results, Buffon suspected that his result was too low, based on assumptions that were likely too simplistic; he later revised his age for the earth, making a number of changes and obtaining a range of results, the highest being about three billion years. However, this highest revised estimate was apparently not known to the outside world until early in the nineteenth century, well more than a decade following his death.

By the mid-nineteenth century, determining the age of the earth had become a much more pressing scientific issue. One reason was to address the question: Had there been enough time for biological evolution to take place to lead to us? Geologists had made extended examinations of surface features around the globe. In these surveys, they had uncovered many complicated (in some cases very complicated!) layered regions (fig. 12.1). Geologists assumed that the layers were arranged by age, with the topmost being the youngest. Why would they have made such an assumption? And from where might the new material that makes up the later layers have come?

In addition, geologists made correlations between types and thicknesses of layers from different geographical regions that enabled them to make reasonable inferences on common dates of features from these different

Figure 12.1. Geological layers in Utah, immediately underlying the Poison Strip Member of the Cedar Mountain Formation. Courtesy of James I. Kirkland, Utah Geological Survey.

geographical regions. Thus, worldwide, geologists had a pretty good idea of the relative ages of the various layered structures and the relations of the ages of strata from different geographical regions. Why should they care about absolute ages? Why weren't relative ages sufficient for all of their purposes (other than curiosity)? One glaring nongeological example comes from the third section of this book, as already noted above: although not so much of direct interest to geologists, biologists wondered whether the earth was old enough for there to have been time for natural selection to have operated to yield the present cornucopia of biological complexity and diversity—another example of the unity of science.

Charles Darwin apparently worried about this issue. He therefore estimated what he thought to be a lower limit to the age of the earth based on his calculation of the erosion time for chalk cliffs in southeast England. Of course, to make such a calculation, Darwin needed to know the rate of erosion. It was clearly too slow for him to measure, so he had to guess. Given his guess of 1 centimeter per century, he calculated that the earth must be at least 306,662,400 years old, a value he published in the first edition of his world-famous *On the Origin of Species*. Given the crudeness, which he doubtless realized, of his value for the erosion rate, why did he publish such a precise number for the earth's age? I know of no explanation that he gave. My own inference—rather, pure guess—is that Darwin gave this precise a number so that anyone who wanted to check on his arithmetic would have a precise number against which to compare their result. This part of Darwin's book received very harsh criticism, primarily because of the lack of support for the rate of erosion that he used. Lo and behold, all of the many future editions of *On the Origin of Species* were published without that result on the age of the earth; it seems to have simply disappeared.

The Attempt to Apply Modern Physics

The major battleground in the mid-nineteenth century about the earth's age was the confrontation between geologists and a physicist. The latter, William Thomson, now known as Lord Kelvin, was a mathematical physicist with an excellent, worldwide reputation. Lord Kelvin decided to take on the age problem, probably due to his other, physics, researches, which indicated that all hot things in the world must eventually cool down. So, Kelvin approached the problem of the age of the earth from this perspective, using quantitative physics, as opposed to the qualitative approach then used by geologists. Geologists of that era knew no way to estimate this age without making, from their surface observations, unjustified and unjustifiable assumptions as did Darwin. Kelvin took two different fundamental approaches, one based on the earth and the other on the sun.

Kelvin assumed that the earth was initially molten, due to the heating when it first formed, via the conversion of gravitational potential energy into heat. Roughly speaking, this heat release is like the heat released when a mass falls and hits the ground: it gets hotter! He thus took as his starting temperature for the interior of the earth the melting temperature of iron, which he assumed was about 5,000 Kelvin, to use modern terms. (This unit of temperature had not yet been created nor, a fortiori, named for Kelvin when he started working on the age of the earth.) Kelvin knew that the melting temperature of iron, or of other materials for that matter, depended on the local pressure, and he knew that the pressure was extremely high at the center of the earth, due to the weight of the earth, but he had no idea of the quantitative effect of pressure on the melting temperature; he therefore assumed that the melting temperature would be the same as in a laboratory on the earth's surface. Kelvin then made use of estimates of how well rocks conduct heat, although he did not even know what kinds of materials were deep inside the earth. In his calculations, he used formulas that describe heat flow, to determine the time it would have taken the earth from a molten state to reach its present temperature near the surface. Given that the earth's surface receives enormous amounts of heat from the sun, he used measurements of temperature inside mines, which he assumed were virtually immune to the effects of the solar input, especially as the temperatures increased with depth. From these calculations, and with some allowance for uncertainties on his assumptions, Kelvin concluded in the early 1860s that the earth was between about forty and one hundred million years old. Many geologists, by contrast, thought that the earth must be much older but were largely cowed by Kelvin's quantitative approach, which they did not know how to check. There were, however, eloquent skeptics (keep in mind, of course, that scientific controversies should be resolved by evidence, not eloquence).

But first, on to the sun: Kelvin also asked how old the sun could be, if it had provided, throughout its life, energy to the earth at its present rate. To answer, he needed to know the output of energy by the sun and the sources of energy to the sun in order to estimate how long a match could hold. How could the present luminosity (rate of energy output) of the sun be determined? He could measure how much light reached the earth and also assume that the sun emitted the same amount in all directions. So, knowing approximately the distance of the earth from the sun, he could easily calculate the total rate of energy output by the sun; in this way, Kelvin could have concluded that the output was about 4×10^{33} erg per second, where an erg is the unit of energy in the so-called centimeter-gram-second system of units. It is hard for you, or anyone else for that matter, to comprehend fully how much energy this really is; I assure you that it is unbelievably enormous compared with any energy that you are familiar with in your everyday life on the earth.

Now for the other side of the issue: What sources of energy feed this luminosity of the sun? Lord Kelvin could think of three of possible relevance: the energy brought by objects from interplanetary space that impact the sun; exothermic chemical reactions in the sun (that is, reactions that give off ["exo"] energy); and shrinkage of the sun, similar to the energy released when a body falls from a height to the surface of the earth or of the sun.

Neither the magnitude nor the time variability of these various sources of solar energy were well known. Nonetheless, Kelvin was undaunted. He made what he thought were reasonable estimates of the uncertainties in his quantitative assumptions and obtained results both for the age of the sun and for how long into the future we might expect the sun to shine at its present brightness. For the former, Kelvin estimated the age to be under one hundred million years, consistent with his equally vague estimates of the age of the earth. For the future of the sun, he foresaw no more than about twenty million years, before the sun would cool noticeably and thereby wreak havoc with life on the earth.

What were the reactions of the geologists to these combined results? As noted above, most geologists did not know what to do with results that tumbled from equations. Nonetheless, many contemporary geologists were profoundly skeptical of Kelvin's results: they seemed far too low, despite geologists' limits falling well short of the defensible quantitative realm. Also, many geologists and others leveled stinging critiques on Kelvin's assumptions.

This controversy on the age of the earth can be encapsulated in the old saw, an irresistible force meeting an immovable body. The controversy simmered during the last third of the nineteenth century. Neither side yielded. Kelvin remained (mostly) the irresistible force and geologists remained the immovable body, as we re-create in part by the following quotations of the time:

Kelvin on Kelvin (1862): "It seems, therefore, on the whole most probable that the sun has not illuminated the earth for 100 million years, and almost certain that he [sic!] has not done so for 500 million years. As for the future, we may say, with equal certainty, that inhabitants of the earth cannot continue to enjoy the light and heat essential to their life, for many million years longer, unless sources now unknown to us are prepared in the great storehouse of creation."

With this last comment, Lord Kelvin was far more prescient than he likely realized.

Thomas Huxley (articulate polymath; main defender of Darwin's theory) on Kelvin (1869): "Mathematics may be compared to a mill of exquisite workmanship, which grinds you stuff of any degree of fineness; but, nevertheless, what you get out depends on what you put in; and as the grandest mill in the world will not extract wheat-flour from peascods, so pages of formulae will not get a definite result out of loose data."

Thomas Chamberlin (American geologist) on Kelvin (1899): "The fascinating impressiveness of rigorous mathematical analyses, with its atmosphere of precision and elegance, should not blind us to the defects of the premises that condition the whole process. There is perhaps no beguilement more insidious and dangerous than an elaborate and elegant mathematical process built upon unfortified premises."

Both of these last two quotations, one each by Huxley and Chamberlin, say elegantly what was said far more succinctly—and often—early in the computer age (mid- to late twentieth century): garbage in, garbage out. Also note that these quotations are utterly devoid of specifics—they are purely ad hominem statements.

Thomas Chamberlin also on Kelvin also in 1899, but relevant to the sun and not the earth: "Is present knowledge relative to the behavior of matter under such extraordinary conditions as obtain in the interior of the sun sufficiently exhaustive to warrant the assertion that no unrecognized sources of heat reside there? What the internal constitution of the add-ins may be is yet an open question. It is not improbable that they are complex organizations in the seats of enormous energies."

This quotation shows Chamberlin to have also been a man with acute foresight.

The Radioactive Clock

Matters remained in this unresolved state until well into the twentieth century, when the developments described in the previous chapter produced the new knowledge that led to the solution of the problem of the age of the earth. Our hero is none other than Ernest Rutherford. In 1904, eight years after the discovery by Becquerel of radioactivity, Rutherford suggested that radioactivity could be used as a clock. How? It is simple in principle: isolate a measured amount of a radioactive element at some instant and measure the amount left at any later instant to determine the total elapsed time between the two instants, which one could calculate knowing the half-life of the radioactive element. As we shall see below, the practice is a lot more complicated than the principle in using radioactivity to determine the age of the earth. But in 1904, Rutherford obtained a first cut at this age by using radium as the relevant radioactive element. His result was a number for the age of a rock—not the age of the earth, which presumably was far older. So how did Rutherford (and his successors) use radioactivity to estimate the age of a rock? The basic idea was to look at an isolated portion of a rock, which had radioactive parent and daughter elements present. The half-life of that radioactive parent was measured separately. The key assumption was that only the parent was present when the rock first formed from its molten predecessor. Another pair of assumptions was that no parent or daughter atoms left or entered the rock after it formed; the rock was truly isolated in these senses from its environment. If true, the age determination would be straightforward.

Rutherford gave a public lecture on his idea and on a preliminary result. In his own words:

> I came into the room, which was half dark, and presently spotted Lord Kelvin in the audience and realized that I was in for trouble at the last part of the speech doing with the age of the earth, where my views conflicted with his. To my relief, Kelvin fell fast asleep, but as I came to the important point, I saw the old bird sit up, open an eye and cock a baleful glance at me! Then a sudden inspiration came, and I said Lord Kelvin had limited the age of the Earth provided no new source of heat was discovered. That prophetic utterance refers to what we are now considering tonight, radium! Behold! The old boy beamed upon me.

Note that only the age of a rock, since it first formed from its presumably liquid predecessor condition, could be so determined. How, though, does one know that the rock whose age is being estimated is as old as the earth? One doesn't! So, what recourse does one have? Should one determine the age of every rock and see which is oldest? Such an approach is clearly not very practical. Moreover, how would we even know that the oldest rock now around has been here since the birth of the earth? We wouldn't. We can thus use radioactive dating of rocks to at best obtain a reliable *lower bound* on the earth's age.

In any event, what are some of the potential pitfalls of this method of determining the age of a rock, let alone that of the earth? First, not all rocks solidified from a liquid form. Second, when it originally formed, a rock may have already had present, for whatever reason, some unknown amount of the daughter element in addition to the parent, radioactive element. Third, the rock may have undergone one or more partial melting episodes since its first formation, with some of the parent and/or the daughter element having escaped, and/or different amounts of either having entered. With these potential problems, how could we then believe any results obtained from the straightforward analysis described above? Redundancy to the (partial) rescue! One could use two (or more) independent parent-daughter pairs that have different susceptibilities to the various possible systematic errors. Then, if consistent results were obtained, one could have more confidence in the reliability of the age determination. Caution: one must use pairs of elements that have half-lives suitable for the problem at hand. Thus, for an extreme example, if one were dating a rock in the billion-year-old range, using a radioactive element whose half-life was of order 10 years, there would very likely be no discernible amount of the parent left from which to determine a reliable age for the rock.

After Rutherford's initial idea for using radioactivity as a clock (shades of Galileo and Io!) became known, the field of dating via radioactivity greatly expanded. By the early 1920s reasonably reliable results were being obtained. The age of the earth was estimated by different workers as being between one billion and eight billion years, with even the lowest value of this range being well more than almost all previous estimates made by other means. Still, there were many skeptics among well-respected geologists. A typical example was F. W. Clarke, of the U.S. Geological Survey, a well-respected organization. He declared in 1924: "From chemical denudation, from paleontological evidence, and from astronomical data, the age of the earth has been fixed with a noteworthy degree of agreement at something between 50 and 150 million years. The high values found by radioactive measurements are therefore to be suspected until the discrepancies shall have been explained." Note the irony: the age range that Clarke here endorses encompasses most of the values that Kelvin had put forth more than a half-century earlier and was hooted down by geologists of that era.

Why, however, were the estimates of the earth's age using radioactive clocks so discrepant with one another, spanning almost a factor of ten in age? The methods of measurement were not so simple to carry out accurately; it took many years to refine these techniques to use radioactivity to accurately estimate the age of rocks. Also, different rocks could and usually did have intrinsically different ages.

Given that we can now use radioactivity to accurately measure the age of rocks, how then do we estimate the actual age of the earth? We search for pristine bodies, those that we believe were around at the formation of the planetary system. Good prospects are meteorites. These are objects that were in orbit around the sun but that collided with the earth and survived, at least in part, the passage through its atmosphere before striking the ground and being recovered. A good place to find them is Antarctica; because of the open white expanse, a meteorite will stick out and be easily spotted for a long time before being buried and no longer easily discovered.

The most modern radioactive dating results, from a suite of different meteorites, have yielded a remarkable concordance, values of about 4.567 billion years, with an uncertainty of the order of 1 million years. This number is easy to remember and fairly reliable as an estimate for the earth's age, based on models of planet formation, which indicate that the planets and the meteorites were essentially cocval (born at the same time).

Here we have a clear triumph of science. The evidence from radioactive dating and redundant decay chains is overwhelming and thus thoroughly convincing to virtually all scientists. Estimates by geologists of the earth's age were totally eclipsed. Rationality triumphed, based on science that has withstood extensive criticism and emerged with clean hands.

Where, specifically, did Lord Kelvin go wrong? He made mutually consistent estimates about the earth's age and the sun's shining. Both were

wrong due in large part to radioactivity and nuclear fusion, respectively, neither of which was known or even dreamed about in the mid-nineteenth century. For the earth, radioactivity provided a continuous source of substantial
heat, and thus the earth did not just cool from an original high temperature
as Kelvin had assumed. For the sun, there is, which we did not discuss, a
plentiful supply of energy from a source that was not even dreamed of—no
background knowledge available!—in Kelvin's day: the fusion of hydrogen
into helium, at the very high temperatures and pressures near the center of
the sun. This process provides huge and nearly inexhaustible supplies of energy (recall that $E = mc^2$ and be advised that a helium nucleus has 0.7 percent less mass than four hydrogen nuclei; this excess mass is converted into
energy in the fusion process). This energy, leaving the sun mostly as light,
will, we estimate, keep the earth warm for more than another five billion
years. So, we personally don't need to worry. Nor do our grandchildren. At
least not on that score.

Finally, we note the (close to) prescience of Ernest Rutherford on this
subject. In 1904, he pointed out: "The discovery of radioactive elements,
which in their disintegration liberate enormous amounts [of energy], thus
increases the possible limits of the duration of life on this planet [via the
sun's radiation], and allows the time claimed by the geologists and biologists
for the process of evolution." He didn't think of fusion, of course, as it lay way
in the future, and a fortiori of its even more enormous releases of energy.
Nor did he know then that the sun was made predominantly of hydrogen,
but he was clearly thinking about this general issue in an important way.

We will next change the subject dramatically and begin our exploration
of the earth's fossils.

THIRTEEN

Fossils

W hat is a fossil? The origin of the word ties to an object "dug up." In
ancient usage, anything interesting qualified. In modern usage, fossils are restricted to things once—or still—connected with organic matter,
although definitely not alive. Footprints, and such, qualify as trace fossils.
What percentage of the living organisms in any era turn into actual fossils?
No one knows. But it is clear that fossils are rare—very rare—compared to
the number of organisms that have ever lived, even though fossils generally
preserve only fragments of organisms. The conditions have to be just right
for fossils to form. Most dead things are almost immediately eaten by scavengers or decay, usually via being attacked by bacteria. To form a fossil, a
newly dead organism probably must be under water, immediately buried by
sediment, and/or be in a very dry or cold region. The fossil record is perforce
also a biased view into previously living organisms. For example, fossil formation favors organisms with hard shells or skeletons; soft organisms decay
too easily. There is, however, at least one exception: organisms preserved in
amber (see below).

How are fossils produced? Count the ways; there are many, and within
each category there are many variations. The word "taphonomy" (alas, jargon!) denotes the study or science of the production of fossils. By way of
introduction to it, let us discuss a few common means of producing fossils.

One such method is permineralization. Minerals, such as silicates,
calcium and magnesium carbonates, and iron sulfides, in the immediate
neighborhood of dead matter can seep within the cells of the deceased

Figure 13.1. A permineralized trilobite (see text). Wikimedia Commons, User: DanielCD.

organism while preserving the cell structure. Amazing detail is often maintained. These minerals replace the cell walls and help to prevent the compaction of the tissues of an organism from the increased pressure due to sediments, which gradually build up over the fossil. A permineralized trilobite is shown here (fig. 13.1). This three-lobed ancient creature, stemming from an origin of its species about a half-billion years ago, had amazing staying power, with fossils from it dating from about 520 to 250 million years ago. Trilobites disappeared in a major extinction event. Major extinctions have been pinpointed by scientists over about the past half-billion or so years of earth history. One such event, about 250 million years ago, was apparently the largest in that it has been (crudely?) estimated that over 70 percent of all of the different types of organisms perished—became extinct—in that event. These extinction events were doubtless not instantaneous but doubtless also extremely short compared to the time intervals between them.

Trilobites were named for their three axial lobes. Sizes of trilobites ranged from about 1 millimeter to 70 centimeters, an astonishing ratio of seven hundred to one, putting to shame the corresponding size ratio of dogs, which we usually think is rather large, ranging from smaller than Pekingese to somewhat larger than Great Danes, which may be separated by a factor of ten or so in length. The fact that trilobites molted as they grew gave rise to the possibility that a given trilobite might have left behind more than one fossil. They are also the earliest organisms suspected to have had vision that covered a full 360 degrees (see the two protrusions in the figure).

Another method of fossil formation was molds and casts, in which an organism with hard parts is buried in sediments and is dissolved, or otherwise destroyed, leaving behind a hole, or mold. This mold, filled with other minerals, then becomes the cast. A beautiful illustration of such a cast is of a 112-million-year-old ammonite (fig. 13.2). These are typically ribbed, spiral-formed shells that existed as far back as 410 million years ago and thrived in the seas from about 250 to 66 million years ago, perishing in the event that did in the dinosaurs. Of course, much more could be said about molds and casts, just as with all topics we have covered and will cover, but we will refrain from so saying.

Resin from some kinds of (ancient) trees, produced by complicated chemical reactions, provides a third, rather unique mode of preservation (fig. 13.3). The hardened resin, called amber, has preserved bacteria,

Figure 13.2. The cast of a 112-million-year-old ammonite. Photo from the Virtual Fossil Museum, CC BY-NC 4.0.

Figure 13.3. An ancient ant preserved in amber. Anders Leth Damgaard, http://www.amber
-inclusions.dk/ CC BY-SA 4.0.

insects, spiders plus webs, flowers, fruit, even dinosaur feathers—but so far
no evidence has been found of preserved DNA, the *Jurassic Park* movies
notwithstanding. Such preservation would constitute a real coup, since the
oldest amber fossils known were formed about 230 million years ago. However, other environmental factors are thought to limit DNA preservation to
far shorter periods of time, perhaps to as little as ten thousand or so years.

With this brief introduction, let us move to a history of humans' understanding of fossils.

The First Scientific Treatments and Discoveries of Fossils

While the ancient Greeks were apparently aware of fossils, describing some
of them as the remains of gigantic animals, the first serious scientific study

of which there is any direct evidence stems from none other than Leonardo da Vinci (1452–1519). As usual for this amazing genius, he was well ahead of his time. How did Leonardo get involved? His contemporaries had noticed the presence on the top of local mountains in Italy of the remains (fossils) of sea animals. How could they have been transported to the tops of mountains? The answer that his contemporaries came up with was simple: the biblical flood was responsible. However, Leonardo, in a detailed discussion of this possibility, showed that explanation to be wanting. In particular, he showed by logical analysis of the various possibilities that there was no way that these sea creatures would have risen with the flood or could have made it on their own from underwater to the mountaintops. He postulated, instead, that the creatures were deposited when the mountaintop was below sea level. That is, rather than the sea creatures going to the mountaintop, the mountaintop took the sea creatures with it from the bottom. Of course, his contemporaries did not accept his prescient conclusion. It is also not clear how familiar contemporary scientists were with his analysis and conclusions. Leonardo did not publish his results. Indeed, the invention of the printing press was then rather recent. Moreover, the notebooks in which Leonardo recorded his scientific, and other, thoughts were written in an unusual manner: they could be read in the ordinary way only by holding them up to a mirror. Why? We can only speculate. My speculation is that Leonardo was left-handed and that, writing in this way with a pen, he could avoid smearing what he had already written!

Finally, I must add, his analysis of the sea creatures' appearance on the tops of mountains was not what I would consider eminently clear writing, at least not in the English translation that I read. But it was an early, and in my view important, contribution to our knowledge of fossils, even though it was likely rather late in coming to the general attention of scientists.

Over the next several centuries, interest in fossils as a worthy subject of scientific inquiry gradually increased. Conrad Gesner (1516–1565), a Swiss scientist interested in fossils, apparently made three main contributions. First, he organized direct communications among scientists, of course via snail mail, to discuss relevant issues. Second, he instituted the idea of including drawings to supplement words in describing objects being studied. And third, he invented a cabinet with drawers in which to store fossils in a systematic way. Now, more than five hundred years later, his cabinet design is still in use with practically no improvements.

I must also add that other sources that I have read do not mention the first and third of these contributions. As I have not seen primary evidence on any of these three contributions, I cannot discount that any of them—or even all of them!—are misleading in some serious way(s) or simply false. Although all accounts seem to agree that Gesner was indeed a very clever person.

As far as I know, the field of fossils went through a fallow period after the passing of Gesner, a period that lasted over two centuries. The next major figure I mention is Georges Cuvier (1769–1832), a very precocious and brilliant Frenchman. He made significant contributions to comparative anatomy, including the reconstruction of whole organisms from the recovery of only a few bones; this area was important to scientists trying to understand how to reconstruct organisms from very limited clues from fossils. Thus, in 1798, at age twenty-nine, Cuvier concluded: "Today comparative anatomy has reached such a point of perfection that, after inspecting a single bone, one can often determine the class, and sometimes even the genus of the animal to which it belonged, above all if that bone belonged to the head or the limbs. . . . This is because the number, direction, and shape of the bones that compose each part of an animal's body are always in a necessary relation to all the other parts, in such a way that—up to a point—one can infer the whole from any one of them and vice versa." (Probably translation from French.)

Cuvier was apparently the first person to argue, with supporting evidence, that the extinction of organisms was a real phenomenon. In particular, when confronted with the fossil of a mammoth, of which no living specimens were known, he drew the correct conclusion that the mammoth was extinct and therefore that the earth can and does undergo dramatic changes. He was a proponent of the catastrophism school of earth history, in contrast to the English school of Charles Lyell, who believed that the earth went through cycles but stayed pretty much the same. Cuvier also helped to establish the field of stratigraphy in which one learns about the past from the layering of the ground. In addition, Cuvier apparently coined the term "paleontology," which is the study of fossils (see, too, above). He also identified, in print, and named the pterodactyl, the first known flying reptile, which was a cousin of dinosaurs from that era. Like most mortals, though, Cuvier was not always prescient nor even correct. I end with one example of the latter. In 1821, he stated, "No larger animals remained undiscovered."

We now switch the scene to England, where that prediction of Cuvier's was soon to be proven incorrect. A young woman named Mary Anning lived in the southwest corner of England near exposed strata that were about two hundred million years old and contained many and varied fossil specimens (fig. 13.4). Having a background of poverty but possessing a keen sense of entrepreneurship and knowledge about her specimens, Anning established a lively business selling many of her fossil finds. In the process she increased— not by accident—the awareness among the citizenry of the existence of fossils of many now extinct organisms, not to mention awareness of her own existence. Her fossil discoveries were extraordinary, and her head for publicity and business acumen lifted her out of poverty; rich she apparently never

Figure 13.4. The plesiosaurus, an organism discovered by Mary Anning in the early 1820s. This creature attained a length of about 3.5 meters. Kim Alaniz, CC BY 2.0.

became. As an indication of her fame, two examples come to mind. First, a statement Lady Harriet Sylvester (the widow of the recorder of London, its chief judge) made in 1824, when Anning was at most twenty-five years old:

> The extraordinary thing in this young woman is that she has made herself so thoroughly acquainted with the science that the moment she finds any bones she knows to what tribe they belong. She fixes the bones on a frame with cement and then makes drawings and has them engraved. . . . It is certainly a wonderful instance of divine favour—that this poor, ignorant girl should be so blessed, for by reading and application she has arrived to that degree of knowledge as to be in the habit of writing and talking with professors and other clever men on the subject, and they all acknowledge that she understands more of the science than anyone else in this kingdom.

The second concerns a famous tongue-twister with which likely you are familiar, even though I strongly suspect that you did not know that Mary Anning was apparently the "she." This tongue-twister was composed, about sixty years after her death, in 1908, by the songwriter Terry Sullivan:

> She sells sea shells on the seashore
> The shells she sells are seashells, I'm sure
> So if she sells seashells on the seashore
> Then I'm sure she sells seashore shells.

Discovery of Dinosaurs

On (and back) to larger discoveries (in two senses!), also in England: in the early 1820s a British physician, Gideon Mantell, with an interest in science, or his wife, Mary, who apparently accompanied him on a house call, discovered a large fossil tooth that resembled an iguana tooth but was about *twenty times* its length. Discovery of more large bones by Gideon soon followed. By the 1840s, a scientist with rather sharp elbows, Richard Owen, apparently tried to claim for himself the discovery of these large fossil bones, whose original owners he named in 1842 as "dinosaurs," meaning "terrible lizards." Gideon Mantell, who tried to maintain his proper place under the sun, was seriously injured in a horse and carriage accident, which thereafter seriously impeded his mobility and led to his premature death, leaving the field open for Richard Owen to exploit, and exploit he did, ignoring Mantell's contributions and publicizing his own. Scientists with impure souls are not, and have not been, quite as uncommon as the popular image would lead one to believe.

The next major event occurred in New Jersey, where William Foulke heard about the discovery on a nearby farm of rather large fossil bones. He went to see these specimens and, with the expert help of Joseph Leidy and some locals, systematically dug in the vicinity where the fossils were found. They thus uncovered the first (nearly) complete skeleton of a dinosaur. This discovery was presented in 1858 to the Philadelphia Academy of Sciences, the nearest large city and one that had such an academy. The proceedings of that meeting in 1858 contain the story of that discovery and of the dinosaur in exhaustive, and exhausting, detail. It presents well the flavor in those times of such publications. The apparent fact that no other dinosaur bone of this species has yet been discovered is understandable—nearly half of all known dinosaur genera are as yet apparently represented by only a single specimen.

From the nineteenth century through the present, the search for, and the publicity surrounding, the discovery of dinosaurs increasingly enchanted the public. Most children, myself included, when told about their existence, found dinosaurs fascinating. Dinosaur fossils have now been identified on all seven continents, including therefore Antarctica. How come Antarctica? This continent clearly has a climate that is not conducive to dinosaurs' thriving. True. But dinosaurs, judging from the fossil record, were dominant creatures from about 230 to about 66 million years ago. During that time, the continents underwent substantial movement. There were apparently also times when it was even warm near the poles. For both reasons, Antarctica, which is a word virtually synonymous with frigid, was thus not inhospitable to dinosaurs during much of the time period that they thrived. Finding their fossils now in Antarctica is another matter.

FOURTEEN

Sauropods

A Comprehensive, yet Superficial, Look

We continue our dinosaur discussion but limit our treatment to a probe of only one type of dinosaur, namely sauropods ("lizard feet"), the largest creatures that ever roamed the land. Even by dinosaur standards, they were enormous (see fig. 14.1, below). We discuss now what we know about sauropods, how we know it (what evidence supports these beliefs), and the serious questions whose answers we still seek.

First, though, we'll spend a few paragraphs on classification, because it is a basic part of biology. With little foreknowledge of what I might call fundamental knowledge of biology, our forebears confronted this bewilderingly diverse subject first by trying to sort out its constituents in a sensible manner. This process was taken up once again in the last century. As a result, there are now two systems in use simultaneously. The old one, familiar to most (domain, kingdom, phylum, class, order, family, genus, species), is a mouthful for sure, in which the first listed is the most general and the last the most specific, thus containing the smallest number of organisms. The much newer system is cladistics, developed first by Willi Hennig (1913–1976); a clade is the primary unit, and each clade contains all the organisms that have a common ancestor, including that common ancestor. In this system, one can have, for example, clades within clades. Neither scheme is quite straightforward in application, given the available data, and details in each are often subject to disputes by the proponents of different views.

The classification of dinosaurs, as of other organisms, can be expressed in either system. Hand in hand with classification goes the discovery of new species or clades. Sometimes the reverse happens and a species is eliminated, as when the brontosaurus was absorbed by apatosaurus, after research showed that they belonged to the same species, thus causing the brontosaurus to be jettisoned by the International Commission on Zoological Nomenclature. However, more recently, a long (about three-hundred-page) paper argued that the brontosaurus was indeed worthy of being called a separate species, with the result that it has been restored to its former status among dinosaur species. Other changes will likely be made as more fossils are found and/or other insights surface that are relevant to the classification issue.

Implied by this resurrection of brontosaurus is that the naming of different groups of organisms, certainly including dinosaurs, is a complicated game needing memorization of many names, often composed of unusual syllables, bestowed according to the finder's considered judgment or whim (their choice). Thus, we largely avoid this numbers/names game, which is too long an adventure with little reward in understanding either the science or the sociology.

The Emergence of the Sauropod

The earliest (oldest) fossil bones that have been classified by paleontologists as clearly belonging to a sauropod are dated from their surrounding strata as being about 210–220 million years old, nearly as old (relatively speaking!) as the oldest fossils of dinosaurs in general. These remains were found in Germany in the late nineteenth century. The oldest sauropod fossils found in North America were a humerus, an ulna, a radius, and some vertebrae; these were dated only in September 2014, yielding an age for the bones of between 155 and 160 million years. The basic discovery of these bones began in the western United States near the end of the nineteenth century; continually ever since, more of these materials have been discovered around the world and dated. One usually finds these bones encased in rock, carefully notes the rock's placement and orientation in space, then carefully cuts out the part(s) of interest, outside the bone limits, and sends it (them) to a lab for final scraping away, and dating, via radioactivity, of the rock, and, most important, for studying of the bones themselves.

THE EARTH AND ITS FOSSILS

The populations of sauropods in different regions of the earth, and their evolutions with time, are unknown. As far as I know, too, almost nothing is known about how their separate biological evolutions proceeded after different populations became isolated on continents with no land connections to other continents. These issues are, however, being actively researched.

The Evolution of a Giant

The size evolution of sauropods is a tricky issue. Since the fossils are relatively few in number, how can we reliably distinguish sauropods that were fully grown at their time of death from those at various stages of incomplete growth? Moreover, how do we know that our samples are representative, sizewise, of the population then alive? That is, how can we distinguish the average size of sauropods in any era from individual and, say, gender variations, which may not be small? Before giving an answer, we note the illustrious history of this question, which led to the formulation of Cope's rule, which, roughly speaking, states that, over time, fully grown individuals of a given species tend to evolve to be larger. Interestingly, though it was named for him, this rule was not formulated by Philadelphian Edward Drinker Cope (1840–1896). This rule is not inviolable, nor does it comment on the limits of growth, which are present, albeit different in different situations.

Over the interval from about two hundred million to one hundred million years ago, the size of sauropods may have tripled, but this is a very tentative conclusion and may well not be reliable. In fact, the largest size reported is based on a discovery written about once by Cope in the late 1880s, and never again. It is credited to his measurements of this apparently spectacularly large fossil find that disappeared strangely about a century ago. However, a very recent scholarly analysis of the situation comes to the (tentative) conclusion that there was a critical typo in Cope's write-up and that this fossil find wasn't uniquely large. Whether this analysis will stand up—or sit down!—remains to be seen.

No one knows for sure what was the impetus, or driving force, for sauropods to become such huge organisms. The largest sauropod discovered up to the fall of 2014 was found in Argentina and was about 30 meters from head to end of tail (fig. 14.1). In August 2017, a substantially larger sauropod

Figure 14.1. The second-largest known sauropod, found in Argentina (see text). From W. I. Sellers, L. Margetts, R. A. Coria, and P. L. Manning, "March of the Titans: The Locomotor Capabilities of Sauropod Dinosaurs," *PLoS ONE* 8, no. 10 (2013): e78733. Image copyright Phillip L. Manning.

was discovered, also in Argentina, and was estimated to have been about 40 meters in total length. Speculations on some of the possible evolutionary benefits of bigness are: ability to cover more ground to find food (some sauropods were likely migratory), improved defense against predators, greater success in predation (although sauropods were herbivores!), more success in mating, increased intelligence (larger brains), increased longevity (not clear why this would be related to size other than increased resistance to predators), increased thermal inertia (better resistance to sudden environmental temperature changes), and increased survival through lean times (not clear this wouldn't be decreased survival: less agile versus dominant in obtaining more of the more limited resources).

On the other side of the coin: What limited the growth of sauropods? Again, no secure knowledge answers this part of the question. Some possible contributing factors, however, are: increased requirements for food and water, with either one or both not easily available; physiology of the bones, hearts, and circulatory systems; an increased susceptibility to extinction (the increased development time of individuals implies a longer time between generations and a consequent slower adaptation to environmental change); lower fecundity; and constraints of a limited area of habitat (for example, an island location). Then there is the issue of cooling, in which the time scale is proportional to the size of the organism. Also, there is the ability to supply blood and oxygen throughout, especially to the brain. Another: the limited height of trees; adding height to the sauropod could then be counterproductive. Of course, we then must understand the limit on tree height. And then there is the unity of science in, for example, the argument from physics of the bone diameter of an animal needed to support its weight.

Recently there have been detailed studies of what are called evolutionary cascade models to explain sauropod gigantism. These models trace various sauropod traits, aspects of their environment, and so on, showing how each leads through its own path to the very high body mass of the sauropod. We omit these details, thus giving only a slight flavor of the overall models.

Subsidiary questions here include: How fast (or, equivalently, how slow!) did sauropods move, and for how long (endurance)? And how do the answers depend on the size of the individual? Basic evidence consists of preserved, bathtub-sized, tracks in rocks made by some individual sauropods. One can sometimes distinguish tracks of juveniles from those of adults (interestingly, sauropod young seemed to preferentially live near shores; see the later discussion of birthplace choices). The front and hind limb tracks imply: (1) an upright—but four-legged—gait, from the relative depth of impressions of feet, thus movement on all fours; and (2) the length of a pace. One article on this issue of speed of locomotion concludes that the sauropods' speed was comparable to that of a walking human. Such studies are still young; applying physics via computer models (more unity of science!) has not yet led to what to me seem to be very reliable, quantitative results. Tune in again in a generation—or maybe sooner.

What (and How) Did a Sauropod Eat?

The evidence is very strong that sauropods were herbivores. There are several lines of evidence leading to this conclusion, the strongest being the fossil teeth. The sauropod had thick, flat teeth, presumably optimized for cutting thick foliage. Sauropods also had teeth of other shapes; all, though, seemed tailored for eating different kinds of plants. Mastication seems to have been largely performed in their gizzards, perhaps using swallowed rocks to so do. Carnivores, by contrast, have some long, pointed teeth to help in tearing at their prey's flesh (please excuse this perhaps too vivid, although probably accurate, imagery).

The fossil record, if properly interpreted, also shows us that each of a sauropod's teeth was replaced by a new tooth about every thirty-five or sixty days, depending, respectively, on whether the sauropod was a low- to mid-canopy or a mid- to high-canopy browser. Why? One reasonable guess is that there was more grit on the lower canopy plant material than on the higher, thus wearing out the sauropod's teeth more quickly. But why such a sharp dichotomy in replacement rate? Why not intermediate replacement rates?

How, though, do we deduce these tooth replacement rates? There are two main parts to the answer. The first is based on the discovery of lines on teeth, named after their discoverer, Victor von Ebner, over a century ago ("lines of von Ebner"). These lines are laid down one per day. How do we know? There are creatures living today, such as crocodiles, that are strongly believed to be genetically related to dinosaurs, and these contemporary creatures show teeth with von Ebner lines that are added, one per day. The second part, or line of evidence, concerns the length of time a tooth is in use. In the mid-1990s, Gregory Erickson noted that dinosaurs had teeth in waiting behind those in use. By counting the von Ebner lines for each tooth in a line, one can check on the consistency of an obvious deduction: a tooth will be in use for about the difference in the number of von Ebner lines present in the tooth in current use minus the number of such lines present at the same time in the tooth next in line. If each tooth is in use for the same amount of time, this difference in von Ebner lines between one tooth and the one it is replaced by will be the same. So, we can check on the consistency of this interpretation by comparing a lot of pairs. With modern, related creatures, we can check on this interpretation of the length of time between tooth replacements, as we have more than a single fossil to deal with; we have generations of related living creatures! And the consistency checks out.

All in all, though not impregnable, this story seems pretty solid: sauropods did change teeth rapidly. But the question then arises: If sauropods had lifetimes of the order of, say, sixty years (see below), why don't we find many, many more of their tooth fossils than we do of the sauropods? Perhaps when discarded, sauropod teeth were virtually all worn out and quickly disintegrated. A premier researcher in this area, Michael D'Emic, suggested to me when I asked, that sauropods may have lived most of their lives in places different from where they died and thus most of their teeth would have been in different places from where sauropod fossils were found. Of course, it is also true that the teeth are rather small and thus more easily escape detection.

From the fossil bones of various individual sauropods, and especially from those that are nearly complete sets of individuals' bones, scientists reconstruct sauropods as having a rather small, and light, head and a very long, and relatively light, neck. Apparently, though, rather few examples of fossils of sauropod heads have been uncovered, perhaps because their lightness made them prone to destruction, especially by predators and scavengers. This structure of the head and neck of sauropods would have allowed these creatures to reach food at the tops of trees (see below) more easily. Obviously, if the head-neck combination were very heavy, it might not have been possible for the organism to lift its head and neck. Examination of the few head and neck fossils makes it clear that the neck bones are rather hollowed out, presumably to make the neck as light as feasible; food (see below) is not masticated much in the mouth to keep that body part small and light as well. The main body, to which the neck is anchored, is rather large, necessary to carry out the main processing of the food intake. We speculate that, because of its overall large size and small mouth, the sauropod may have spent a major fraction of its waking hours eating, especially when it was very young and gaining weight very rapidly (see below). The sauropod also had a quite long tail; some people — but only some! — inferred that this tail helped sauropods with balance when they were eating from the tops of trees.

Herbivores, such as sauropods, of course ate only vegetation. We infer, from their large sizes and long necks, that they ate primarily leaves from the tops of contemporary trees. Did they primarily move their heads sideways as they ate their way through the available leaves, or did they move their heads vertically, too? We do not know but suspect that they moved their heads vertically as little as possible given that such movement uses up much more energy, an especially important consideration given the large fraction of time that sauropods may have spent in eating. Was there an evolutionary push for the length of the sauropod neck to increase to allow it to reach the top of the vegetation supply? Did the height of then contemporary trees increase first? The answers are, respectively, a tentative yes and a less tentative yes. Another thought proposed by others is that at least some sauropods may have foraged for food like present-day moose, using their long necks near shore to reach the bottom of lakes for vegetation. Many unknowns—thus a lot more to discover, and this field of research is very active.

Breath, Blood, and Body Temperature

Breathing with a diaphragm arrangement like ours might have been rather difficult for a sauropod, given their very long necks. It is thus plausible, and the belief of many who studied this problem, that sauropods breathed with the technique used by birds, whose ancestors were dinosaurs. The avian system has multiple air sacs. The apparently low value of the percentage of oxygen then in the atmosphere implies that dinosaurs had an advantage over mammals, whose breathing relied on a diaphragm system. Or, perhaps by then, oxygen levels were much higher. These are only speculations; there is no reliable evidence of which I am aware.

Understanding the circulatory system of sauropods poses perhaps even more difficult problems: there is a *huge* pressure needed to pump blood in the manner of mammals, about six or so times what we need, given the enormous height of many sauropods. Many possible solutions have been suggested; one involves a lowering of the metabolism rate as the sauropod grew, which implies a need for a heart of much smaller mass. Extrapolating directly from our situation implies that a sauropod's heart would have weighed a mere 7 tons or so! Another solution of nature's was in order but is not now known, only speculated about.

What was the body temperature of sauropods? And how did they control this temperature, and within what ranges? The answers may not be simple: some organisms may control body temperature by a mixture of techniques. Perhaps the mixture depended on age. There is mostly only speculation at this stage available to address this issue of temperature in sauropods. However, an ingenious idea (another example of the unity of science) involves attempting to determine temperature via isotope measurements. How does this work? There is a chemical in bones called bioapatite that appears to have good potential for use as a proxy for the temperature of the host organism. There is a preference for certain isotopes of carbon and oxygen, two of the elements in bioapatite, to bond with each other in the crystal lattice of this chemical. The key point is that the extent of this bonding is temperature dependent, thus providing the proxy. Can this dependence be reliably measured and calibrated? It can be, and was, done, for example, using contemporary creatures and twelve-million-year-old fossils, although not easily. With these results in hand, the method was used on sauropod fossil bones, yielding temperatures of 36°–38° C, similar to those of most modern mammals, including us at 37° C. How sauropods controlled this temperature, we have only speculation as far as I am aware. Since the sauropod's size changed dramatically as it grew (see below), it may have been that sauropods used different methods of temperature control in different stages of their lives or at least different mixtures of such techniques—speculation squared. So, what is the moral to this inconclusive story? Answer: science usually advances incrementally, often with steps forward being accompanied by steps backward; herein a shiny example of a very innovative approach that may—or may not—prove to be an important step forward in the long run.

We don't know much about the diseases to which sauropods were prone, though most of the (scant) evidence seems to be about cancers. A paper appeared in August 2020 giving very strong evidence of advanced osteosarcoma in a seventy-five-million-year-old horned, herbivore dinosaur, a cancer that afflicts humans usually in the second or third decades of their lives. There are also some very recent studies which claim that dinosaurs suffered from other diseases as well; however, I do not believe that these results have yet been verified, so we won't pursue them here.

Protection from Predators

How did sauropods protect themselves from predators? Active protection might have been twofold: the tail may have been used directly as a weapon and also might have been snapped like a whip to make a very scary noise. Both actions, especially the second, must be registered as bona fide speculations. There is more to say. The toughness of their outer layer may have afforded some protection. Although their head being small, and their neck being thin and thus fragile, may have made them vulnerable to attacks on this part of their body, it was high enough from the ground to make it difficult to reach by contemporary creatures except for large carnivores among the dinosaurs, which may have constituted the most successful predators. I am, however, unaware of any fossil bones of sauropods that show teeth marks of a carnivorous dinosaur made when a sauropod was alive.

The Sauropod Business of Being Born

The life cycle of sauropods is a subject of present intensive study and for which some evidence exists. Just recently someone had the idea to use von Ebner lines on a sauropod fetus to estimate the sauropod's gestation period. Of course, this period would only be a lower-bound estimate because, for example, we don't know how old the fetus was when the first von Ebner line was set down. But making reasonable estimates leads to the conclusion that the gestation period of the egg was likely about six months, although others think this is far too long for an egg to hatch. So, no conclusion yet.

In Argentina a few years back, a large clutch of fossilized eggs was found (fig. 14.2). Individual eggs are not nearly spherical; they appear flattened. The largest diameter is about 10 centimeters. Interestingly, the shape of each egg is different from what we are now used to seeing; today's eggs are mostly oblate, elongated along the polar direction, and asymmetrically so (different shapes near the two poles). Could it be that these sauropod eggs were flattened by the environment, before discovery but after being laid? I don't know, but I doubt that some of them at least wouldn't have then cracked, unless the flattening was, say, very slow. To throw even more

Figure 14.2. A clutch of sauropod eggs from a find in Argentina. Credit: Sinclair Stammers / Science Photo Library.

possible complications into our pot, I add that there is evidence that some dinosaurs laid eggs with leathery, rather than hard and brittle, shells. Thus, they would resemble eggs now laid by some species of tortoises. In the last century dinosaur eggs were discovered in China that closely resembled modern birds' eggs. The mysteries thus expand and deepen.

Also found was a remarkable egg nearly ready to hatch. In an X-ray photograph of this egg we can see clearly the bones of the hatchling as they were disported inside the egg, although it looks as if they were rather disconnected, one from the other, not making for a viable organism (fig. 14.3). Nevertheless, a remarkable remain. It is consistent with a symmetric flattened shape of the egg (squashed down, top to bottom).

What about parental care exercised by the mother in the early childhood of her offspring? From the numbers of eggs apparently laid by one sauropod,

Figure 14.3. The inside of a dinosaur egg. The ratio of the length of an adult sauropod to that at birth was usually about 10 times the ratio for humans, which is about 3.5. Photograph by Diane Scott.

and the apparent clumsiness of an adult sauropod for taking care of small ones, I infer that not much care was likely lavished by an adult sauropod on her offspring, save for standing by to ward off predators and for providing food. We also infer that each newborn, estimated primarily from the egg size to have had a weight of well under 10 kilograms, must have grown extremely rapidly to have survived to reach adulthood, given the likely predator environment and the perhaps unlikelihood that the mother stood by for years. Note that a sauropod adult increased its weight from birth by a factor of about 10,000 or so; in stark contrast, a human adult undergoes from birth an increase in weight by a factor of only 20 or, at most, about 30 or so. For sauropods, the longest dimension increased by a factor of about 35 to as much as a couple of hundred between birth and full adult size for the largest of them (say, 15 centimeters to 40 meters); by contrast, our growth from birth to full adult represents a corresponding increase by a factor of about 3.5. Don't be surprised, though, if these stories on sizes change substantially by the time today's kindergarteners study dinosaurs, as new information continues to be uncovered, which may change the statistics.

The lengths of time for a sauropod to reach sexual maturity, to reach full adult size, and to reach typical old age are hotly debated numbers. They are estimated via sophisticated histological (related to tissues) analysis of fossil bones, which preserve the details of some of the living organism. Whereas these bones do not seem to be built exactly like trees with their tree rings, recent studies have been able to discern stages of bone growth and to estimate their corresponding ages. Although the numbers are still uncertain, current thought puts sexual maturity at an age of ten to twenty years, with full size being reached at about an age of thirty to forty years, and with the lifetime extending perhaps as long as one hundred years but on average more likely to be about sixty years or so; there is no definitive evidence on this issue as no "very old" sauropod fossils have yet been discovered. These numbers, as you can infer, are not too different from those of many contemporary animals and of us. I know of no available information whatever on the evolution of these ages over time, for example, over millions of years.

Although many books have been written about sauropods, we will now abandon these fascinating creatures, noting only (once again!) that they still hold the all-time record for size of biological fauna found on land and that they have left us with many mysteries to solve. New ones are being added all the time. The latest I am aware of is the discovery of a fossil of a dwarf (cow-sized) sauropod found on an ancient island in Romania.

Next up is our discussion of the demise of all dinosaurs along with that of many other organisms.

FIFTEEN

The Demise of the Dinosaurs

Almost everybody knows that these days there are no dinosaurs roaming the earth. In fact, looking at fossils throughout the world, scientists concluded that dinosaurs disappeared close to 66 million years ago. They seem to have been the dominant land creatures for about the previous 150 million years. Why did they all disappear, apparently rather suddenly? Such questions about the demise of dinosaurs have percolated through the scientific and lay communities alike just about since the discovery of the first dinosaur fossils. Many ideas, or hypotheses, had been offered in the past, none with convincing evidence. These included, but were by no means limited to, outpourings from a nearby supernova, deadly disease, rapid change in the environment disastrously interrupting or changing food supplies, and the rise of mammals who feasted on dinosaur eggs. Finding the answer to this question about the demise of dinosaurs is a detective story for the ages to which we now turn.

New Model(s) of Dinosaurs' Disappearance

In the late 1970s, Walter Alvarez, then at Columbia University, now at the University of California Berkeley, together with a colleague, Bill Lowrie, also a geologist, sought to use paleomagnetism to infer tectonic motions in Italy. They sought to learn about the rotation of Italy's continental crust (microplate) over the past one hundred million years. They proposed to reconstruct this history by finding samples of sedimentary rocks of various ages and determining, in a laboratory setting, the corresponding directions of the rocks' magnetic field through their detrital remanent magnetism (DRM; when magnetic grains are deposited in forming sedimentary rocks, they align with the earth's field [recall chapter 10 and the related phenomenon of TRM]). In this way, they could trace out the motions of Italy relative to the geographic poles, because the positions of the earth's magnetic pole were thought to have been always near a geographic pole, except when reversing. So off they went to Italy to collect the necessary samples.

In Gubbio (about 200 kilometers north of Rome and reasonably near Perugia), strata were exposed reaching back on the order of one hundred million years. Unfortunately, Alvarez and Lowrie discovered that their original goal was not attainable because of other deformations: local conditions had caused obscuring twists of the various strata. Then they realized that some of their samples had DRMs that pointed in nearly the exact opposite direction to the north geographic pole. They had rediscovered reversals of the earth's magnetic field: north pole becoming south pole and vice versa. They thus pursued this finding for the full exposed strata of the past one hundred million years and determined with reasonable accuracy the history of the magnetic field flips of the earth during that period—a more important result than any they would likely have adduced from the realization of their original goal. The deduction of reversals matched those from elsewhere. Only some years later did dating via radioactivity show that the dates of the flips matched well, as well. End of story? Hardly.

During the period of collecting their samples of rocks, Walter Alvarez became intrigued by a very thin layer of clay—hereafter the KT layer—that had been pointed out to him by an Italian colleague, Isabella Premoli Silva, as marking a sharp boundary between two geologic eras, the Kretaceous (English—Cretaceous, older) and the Tertiary (younger). This layer was very unlike its nearest neighbors. Microfossils as were in the layer below were not present in the KT layer. It was also only about 1 centimeter thick, far thinner than adjacent layers. His former professor at Princeton, Al Fischer, later stressed, in a lecture that Walter had invited him to give at Columbia, that this strange KT layer with no microfossils dated at least approximately from the time of the extinction of dinosaurs. Alvarez soon thereafter decided that the problem posed by this latter extinction might be approached through study of the KT layer. Having the mind that Louis Pasteur referred to, Alvarez took a sample of the KT layer back with him to the United States, along with samples of the layers directly above, directly below, and far below. Although the Italian rock layers were twisted over time in ways that made the original goal of their research unattainable, as I have mentioned, this unusual layer and its neighbors, as we'll see, proved extremely fruitful—serendipity in action!

The age of the KT layer was about 66 million years, and part of the younger era, starting at this boundary, was in 2008 renamed the Paleogene period—courtesy of the International Commission on Stratigraphy, whose work underlay this change. Also, as noted above, this boundary was known to mark the time, at least roughly, when the dinosaurs and other animals and plants disappeared from the earth. The plot thickens over this thin layer! The limestone bed below, as expected, had some clay, but was mainly calcium carbonate mostly from the fossils of foraminifera, or forams for short. This material is commonly called limestone and comes mainly from skeletons of previously living organisms, such as forams; limestone, for example, formed the building blocks of the Egyptian pyramids. Judging from the fossil records, scientists inferred that these forams inhabited the waters of the earth from about 540 million years ago until today, although some species of forams perished along the way. Their sizes varied from less than 1 millimeter in length to a maximum of about 20 centimeters, with the smaller sizes being by far the more numerous (fig. 15.1); they are found on the sea floor.

Back to the layers: by contrast, the KT layer contained primarily clay with virtually no calcium carbonate and no forams. The layer directly above again was almost normal, being primarily calcium carbonate, with different kinds of, and much smaller, forams. What happened in the formation of this thin middle layer that made it so different? It was an intriguing question. As noted, Alvarez decided to pursue it because of its possible connection with the disappearance of dinosaurs, a major problem in search of a convincing solution. How could this question be addressed? Walter discussed this issue with his father, the famous physicist Luis Alvarez. They decided that there were two main possibilities. The first was that the thin layer had been deposited at the normal rate, but for some reason neither calcium carbonate nor

Figure 15.1. Various foraminifera. The true size of the figure, side to side, is about 5.5 millimeters. Photo by Alain Couette, http://www.arenophile.fr/Pages_IMG/P991d.html, CC BY-SA 3.0.

forams were around, presumably due to the extinction of the latter whose mineral remains primarily form the former. The second was that the thin layer had been deposited very quickly, for some as yet unknown reason, with no time for normal calcium carbonate deposits to have been made.

How could these two possibilities—the normal rate and the fast rate of deposit—be distinguished? What was needed was a "judge," a substance also deposited but known to be deposited *at a certain constant rate.* Maybe there was some other constituent of the strata that was known, from other evidence, to be deposited at a constant rate. Determining its concentration in the layer would allow the two possibilities to be distinguished: a normal amount (and one could calculate what this amount would be) would indicate a normal rate of deposit, whereas a small amount would indicate a rapid deposition of this 1-centimeter-thick layer.

Luis suggested: Why not use ^{10}Be, a radioactive isotope of the element beryllium, as this proxy (unity of science!)? This isotope was known to be produced by very high energy—that is, very high velocity—cosmic rays striking oxygen and nitrogen atoms in our atmosphere. (Cosmic rays were named when discovered around 1912 before it was realized that they are really particles, mostly protons, that come from the cosmos somewhere— still a subject of somewhat contentious study.) A textbook noted that the half-life of this radioactive isotope was 2.5 million years. This half-life was long enough to allow a useful determination of the concentration of ^{10}Be after about sixty-six million years, given the very sensitive mass spectrometers then available. (These spectrometers are instruments that separate the particle contents of a material based on their differences in mass, just what is needed to separate isotopes.) Working backward from ^{10}Be's known half-life, one could infer the concentration of the ^{10}Be that was in the clay layer at its formation and thus distinguish between the fast and the slow deposition. There was, however, a bitter pill in the offing: no ^{10}Be could be detected! Why? It turned out that the value given in the textbook for its half-life was wrong; the correct value was only 1.5 million years. Trying to use this proxy therefore became hopeless because the half-life was so short that not enough ^{10}Be was left after sixty-six million years to then be measurable.

Undaunted, Luis Alvarez soon recalled that the element iridium was naturally very scarce on the surface of the earth. Because of its high density, and primarily its dissolvability in molten iron, it had apparently mostly settled to the earth's core with the iron when the earth was—near its beginning—in molten form. But if iridium has fallen on the earth over the last approximately sixty-six million years at a rather constant rate from dust in interplanetary space, then he calculated that were the KT layer deposited at a normal rate, the concentration of iridium would have had a value of about 0.1 part per billion. That is, on average, one particle of every ten billion would have been iridium. However, if this layer had been deposited quickly, the concentration of iridium would have been well below detectability. The ability to detect iridium at the level of 0.1 part per billion was just within reach of the then state of the art. So, the Alvarezes approached Frank Asaro, who oversaw the facility at Berkeley that measured very small concentrations of elements via the technique of neutron activation analysis. Neutron activation analysis is a technique in which high-energy neutrons are generated, say in a nuclear reactor, and then used to bombard a sample, making it radioactive. In such arrangements, the radioactivity leads to very high energy light rays coming out of the particles of the sample, each with the distinguishing signature of the element from which it is emitted.

The task given to Frank Asaro was extremely difficult and required painstaking work; because he was quite fully occupied by other, paid-for, work, nine months elapsed before Asaro obtained a result for the iridium in which he had confidence. That result startled the Alvarezes: the concentration of iridium was at the level of 9 parts per billion, about *ninety times higher* than expected for a normal rate of deposition! What could possibly be responsible?

Luis Alvarez thought about a theory proposed less than a decade earlier for the dinosaur demise: two astronomers had proposed that the dinosaurs were wiped out due to the effects of a supernova explosion nearby in the galaxy. If that were true, Luis theorized, then there would be other indications through enhanced concentration of, for example, plutonium, ^{244}Pu, which would have also been produced in the supernova explosion and rained down on the earth along with the iridium. It had a half-life of eighty-three million years and should be easy to detect if the theories of its production in a (nearby) supernova were anywhere near reliable. So, Helen Michael, an expert plutonium chemist at Berkeley, was called upon to work with Frank Asaro to look in the thin clay sample for evidence of this plutonium isotope. This determination, too, was a delicate operation. After completing this stressful work—it required attention virtually nonstop for about twenty-four hours—they reported to the Alvarezes that they had found ^{244}Pu! Amazing; they had now (nearly) proven that the dinosaurs had been done in by a nearby supernova explosion! They could already see the headlines.

Frank Asaro and Walter Alvarez went to tell this good news to the deputy laboratory director and to discuss what to do before going public. He advised caution: do the entire analysis again, starting directly from the sample, and see whether the same results are obtained. The Alvarezes followed this advice and were very, very glad that they did: the second time around, Michael and Asaro found not a trace of the presence of ^{244}Pu. A positive detection could be attributed to contamination of some sort, but not a negative result. Theory thus dead on arrival. So ended the idea that dinosaurs disappeared in the aftermath of a nearby supernova explosion. Well, if not a supernova, what could be the cause of the increased concentration of iridium? A new idea was needed.

In the 1970s, various people were interested in impacts of space objects on the earth. Walter Alvarez, too, thought of such impacts at the start, but he thought that they would have only local effects; that is why he preferred the idea of a supernova being the cause, as it would necessarily have global effects. Now, given the negative result on plutonium, the idea of an impact by an asteroid or a comet on the earth seemed more appealing. Luis thought that the debris from such an impact would spread globally and wipe out many species of organisms. He calculated that were it an asteroid, it would have had to be about 10 kilometers in diameter to lead to the measured iridium concentration, were it spread worldwide. From study of meteorites, scientists already knew that the concentration therein of iridium was substantially higher than in the earth's crust. So, how impressive would such an impact be in terms of energy released? Very impressive! For a 10-kilometer-diameter asteroid, with a density of about 3 grams per cubic centimeter (for comparison, water has a density of 1 gram per cubic centimeter) and an impact speed of 20 kilometers per second (relatively modest for such a collision), the kinetic energy (energy of motion) released would be equivalent to an explosion of about 10^{14} tons of TNT or about two million times greater than the largest bomb explosion produced on the earth, and about four hundred times larger than was estimated to have been released in the largest known volcanic explosion. Luckily for us, such impacts are not exactly everyday occurrences; their frequency is hard to estimate, but of order once per one hundred million years is a not unfair guess (note the double negative!), although they are not expected by most scientists to be periodic or even close to periodic.

Back to the iridium issue: if the collision model were correct, the iridium in the asteroid (or comet), as assumed in the calculation described above, should have been deposited virtually worldwide. Due to the enormousness of the collision, the resultant debris would have been spread around the earth. In several places suitable strata of roughly sixty-six million years in age were located and samples taken to seek evidence for enhanced iridium concentration. As opposed to the results from the plutonium search, excess iridium did turn up in strata of the correct age at various places, including Denmark, Spain, Colorado, and New Zealand, thus reasonably well, certainly plausibly, establishing the worldwide nature of the results from this postulated impact of an asteroid or a comet. Further, microtektites—small, lightly colored glass objects mostly of millimeter sizes—from that era were found and are known to be produced under conditions of extreme heat, as would accompany such a putative impact. The evidence gathers!

The crater created by this impact should be about 200 kilometers in diameter, judging from theoretical studies (and much smaller, in all variables, experiments), and be about sixty-six million years old. Where was it? Such a crater would be the smoking gun, or at least the gun, if found. Of course, if the impact had been on material that was subducted back into the mantle during the past sixty-six million years, there would be no crater to provide the gun. But the probability of such subduction was not high, given the rather small fraction of the earth's surface that would have been subducted in the last sixty-six million years.

The search for the crater proceeded on several fronts. On one front, in Texas near the Gulf of Mexico, evidence was found of a tsunami—a huge, high sea wave caused by a major disturbance of some sort—dating from that era. This discovery was made in the late 1980s with inferences from it leading to the Yucatán Peninsula as the possible location of the sought-for crater. In 1991, a circular pattern of gravity anomalies (unexpected, or unusual, patterns in the earth's gravitational field) was detected on the Yucatán Peninsula with the suitable diameter, about 200 kilometers. The implication: a crater (under) there was the cause of the gravity anomalies.

As it turned out, two geologists working for Pemex, the national oil company of Mexico, had discovered this crater, named Chicxulub after its location, a bit over a decade earlier. They had told their superiors at Pemex about their having found the world's largest impact crater; their superiors thought it was of volcanic origin and so of no particular interest. Nonetheless, the geologists did discuss the discovery, based on gravity measurements, which

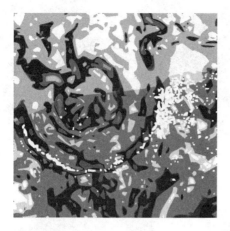

Figure 15.2. Anomalies in the gravitational field in the Yucatán Peninsula (see Figure 15.3), indicated by the circularlike patterns seen. Map by Milan Studio.

disclosed this ring about 200 hundred kilometers in diameter, at a meeting of petroleum geologists in 1981 that was also attended by people discussing the asteroid (or comet) impact linked to the dinosaur demise (fig. 15.2). But this meeting, as many of its kind, was a big one with many parallel sessions and the two groups were unaware of each other's existence, thus postponing by a *decade* the recognition of the crater and its location (fig. 15.3).

Not only was the size of the crater the expected size, but its age, determined from radioactive dating, for example, of the nearby tektites presumably formed by the impact, was also as expected, about sixty-six million years. Case closed? Hardly. That an asteroid or comet hit the earth about sixty-six million years ago, based as well on other radioactivity dating, is a quite robust conclusion. The connection with the dinosaur demise, as well as that of many other species, was less clear. Near coincidence in time does not necessarily imply cause and effect.

Model(s) of Extinction

Rather quickly scientists applied to the dinosaur extinction problem earlier ideas about the aftermath of an atomic bomb war: perhaps a nuclear winter wiped out the dinosaurs. In brief, the idea is that the impact released a prodigious amount of energy and all sorts of debris traveled around the world, igniting horrific fires on all continents, as the debris heated up on reentry into the atmosphere. From these fires, where plant life had existed,

Figure 15.3. Region near the Chicxulub crater, centered at the dark circle's center, right at the coast line of the Yucatán Peninsula. Courtesy of David Shapiro.

the atmosphere was loaded with dust that, after the fires subsided, cut off the warmth of sunlight from the ground for an extended time—the sunlight was reflected by the dust back into space. Consequences for life on the earth? Mass extinctions of species. This model is largely based on numerical simulations; there was then little in the way of other supporting evidence.

Some scientists, on the other hand, especially many of those who had spent much of their professional lives studying extinctions of species, did not simply accept the Alvarezes' theory of this extinction. Over a period of at least a decade, huge numbers of articles were published pointing out their authors' views as to why the impact could not be the cause of the extinctions. For example, it was claimed that fossils of dinosaurs were spread in time over the order of one million years after the impact. The Alvarezes and their supporters answered all of these criticisms with counter papers. A curious question, however, largely remained: How did so many species, especially birds, manage to survive this catastrophe? Where would they hide, and how would they obtain food during a nuclear winter? As far as I know, there is as yet no very satisfactory answer to this question, although we can understand why the large dinosaurs might well have been the most vulnerable: the bigger they are, the harder they fall, as the saying goes.

How all of these other creatures survived is not overly clear; that *is* clear. This is *not* to say that models of survival are lacking; they are plentiful. We simply do not know which, if any, are correct. I will present just one, of necessity in incomplete form, which seems reasonable and was developed by my aforementioned former student Douglas Robertson and his colleagues. They postulate, in effect, two phases of extinction. The first occurred virtually immediately and was due to an intense heat pulse (mostly infrared radiation) that would accompany the worldwide reentry into the atmosphere of ejecta thrown up from the impact. This intense heat, estimated to last for a few hours, would have been the death knell for organisms exposed on the earth's surface. Survivors could have been those shielded from the pulse, such as (small) creatures, including birds, in burrows and caves, and perhaps some under water for extended periods of time (note that this statement is purely qualitative). More generally, it is conceivable that the global effects were nonuniform and that some parts of the globe were partially spared. Almost all these arguments are of course controversial to varying degrees.

The hypothesized second phase of extinctions involved marine animals; in deep, oceanic waters, the heat pulse would not have been immediately lethal but dust and soot from the impact would have destroyed phytoplankton near the surface, dramatically decreasing the food supply of ocean dwellers, as well as decreasing the oxygen supply. Freshwater denizens could have suffered less, because, for example, of groundwater movements circulating edible materials from the bottom. Details here may seriously complicate such simple descriptions and could lead to different end results. In general, be wary of simple-sounding arguments as convincing explanations of complicated situations! Maybe someday . . .

One more point: as far as paleontologists know, there were five other (earlier) major extinctions undergone by life in the past half-billion years or so. But none of these others has been pinned on an impact from an asteroid or comet; maybe one of the same (other) cause(s) of these earlier extinctions was the culprit or partially the culprit about 66 million years ago, too.

This issue of selective survival and causes of earlier extinctions aside, is there any other model afloat to possibly compete with this impact theory of dinosaur extinction? Yes. There were in India about sixty-six million years ago enormous volcanic eruptions that still cover with lava a region about 500,000 square kilometers in area (fig. 15.4). Some scientists have estimated that the original eruptions of lava about sixty-six million years ago may have covered about three times this area, with two-thirds of it having since been eroded away. To put this huge (pre-eroded) area in perspective, consider that it represents a relatively small fraction of the earth's total land area, only about 1 percent.

These lava-covered areas are the so-called Deccan Traps. For those with a bent toward etymology: "Deccan," stemming from Sanskrit, means south, as here in the south part of India, and "traps" means "stairs," from the Swedish; as can be seen in the illustration, the traps resemble staircases, albeit only vaguely. (Why this combination of languages, by the way, is unclear to me.)

Figure 15.4. Part of the Deccan Traps in India, resulting from many episodes of volcanism, and now still covering, after significant erosion, about 500,000 square kilometers in area. Photo courtesy of Mark Richards.

These eruptions would have spewed into the atmosphere serious amounts of sulfur-bearing molecules, mainly sulfur dioxide, inimical to one's health. There is also the possibility of emissions of large amounts of carbon dioxide and its effects on global warming and possibly mass extinctions . . . . A key question is the extent of time over which these eruptions took place. If rapidly, mass extinctions could have occurred. If, instead, the volcanic eruptions were spread more or less evenly over tens of thousands of years, organisms could well survive and populations recover in between the eruptions. So, this key question is converted to one about the dating, specifically the accurate relative dating, of the eruptions in the series that formed the Deccan Traps. Alas, it was not within the grasp of technology, until the past few years, to reliably estimate the ages of these lava beds to better than about 2 percent, not nearly accurate enough to make the needed assessment of the temporal spacing of the eruptions that formed the Deccan Traps. The most recent accuracy achieved, about tenfold better, is still not good enough; see below.

Within about the past five or so years, there has been renewed interest in the Deccan Traps' possible contributions to the extinctions. A group apparently led by Mark Richards, and including Paul Renne and Walter Alvarez, all at the University of California Berkeley, has been investigating the possible role in the dinosaur demise of the Deccan Traps, following previous such investigations by Vincent Courtillot and colleagues in Europe. Of course, the extinctions could be due to some combination of the impact and the Deccan Traps.

Penultimately on this subject, I point out a speculation in answer to the following question: Could there be a causal connection between the impact in the Yucatán and the Deccan Traps in India, or is their (near) temporal coincidence merely a coincidence? The Deccan Traps, new improved radioactive dating discloses, started well before the Chicxulub impact and continued well after it. However, it appears with new more accurate dating, and better understanding of the size of the individual sets of eruptions, that the largest eruption by far took place, as near as we can now tell, at the time of the Chicxulub impact. Radioactivity dating via the argon-argon method, which we will not here explore, has been improved and given results from about sixty-six million years ago with uncertainties at the level of one hundred thousand years—that is, at the 0.2 percent level. This level is quite impressive, but not in my view nearly definitive evidence of the postulated connection.

What could be the mechanism for a noncoincidental connection? The seismic energy delivered to the earth by this impact was likely to have been sizable enough to cause significant displacements on the Indian subcontinent, perhaps enough to have set off volcanoes already near the tipping point. Indeed, careful calculations indicate that this scenario makes some sense; seismically caused displacements there were likely large enough—of the order of a few meters. It is also possible that a different interaction may have been responsible, perhaps the heating due to the dissipation of seismic energy serving to set off volcanism. Although the last relevant word on mechanism has doubtless not yet been said on this subject, it is at least conceivable that the impact could have played a role in one of the major volcanic eruptions that formed the Deccan Traps. Thus, it might be that the impact led to more intense volcanism that led to larger and more rapid releases of poison gases and, say, carbon dioxide, which caused more extinctions than otherwise would have been the case. This type of possibility is being actively investigated by Paul Renne, Mark Richards, and their colleagues.

As for the big question, which of the two was directly responsible for how much of the mass extinction, it will likely not be settled to (nearly) everyone's satisfaction any time in the near future. One new input, however, is a study led by Pincelli Hull from Yale University, that examined in detail sediments from this period of time from under the Atlantic Ocean. She and her group concluded from their analyses that the impact caused the mass extinctions.

A find from about a decade ago, which will be under study for at least another decade, is in the Hell Creek formation in North Dakota and was uncovered by paleontologist Robert DePalma. This site preserves evidence from about ten minutes after the impact at Chicxulub, the time it took seismic waves to propagate from the impact to this site. The site preserves an incredible array of materials. Samples include remains of some dinosaurs, parts in amber; many fish, some of which swallowed microtektites, in abundance at this site; eggs with embryo fossils; and drowned ant nests with ants inside. We don't have space to discuss these in detail. Besides, the discoveries and the analyses are happening rapidly, making it hard to keep current. The possibilities for future discoveries here are, however, definitely exciting.

Another recent work, by Amir Siraj and Avi Loeb at Harvard, developed an interesting case that the impactor was a chunk from a long-period comet rather than from an asteroid.

We now leave our detective story having made a lot of progress but not having definitively and completely solved the mystery of the dinosaurs' demise. Nevertheless, it is probably fair to say that at present the Chicxulub impact is viewed by most scientists as the main, if not the only, cause of the great extinction event of about sixty-six million years ago.

This situation aside, there are other views of the dinosaurs being done in, related to relatively recent and popular societal themes; one is illustrated here (fig. 15.5).

Figure 15.5. The dinosaurs' situation just before their demise; the English words are a recent addition. © 2009 Ted Rall, All Rights Reserved. www.RALL.com.

The Story of Life

SIXTEEN

Story of Life

Darwin's, Wallace's, and Mendel's Contributions

We now embark on the third and last part of the book, the story of life. We have already discussed biological life in the sense of its deceased versions, mostly very long deceased creatures. Here we pick up the story with a new slant: How can we understand life? It is a staggeringly difficult goal to reach, and we are nowhere near reaching it. But in the past hundred and fifty years, we have made enormous progress, some highlights of which we'll discuss. Most likely you have already been exposed to at least some of these highlights in high-school biology. Since not everyone may have learned about them, I will not skip these developments here. Anyway, as a wise professor of mine was prone to say: "Never underestimate the pleasure an audience experiences in being told something it already knows."

Further, as we emphasized in the second part of the book, life has had major influences on this planet, as well as, of course, vice versa. Illustrative are the profound changes in the earth's atmosphere due to life—we cannot understand either in isolation; their intertwined relations in a sense dramatizes the unity of science, the tight connection here being one example of the overall inseparability of the different fields of science—of the interconnections that, together, lead in major part to the title of this book.

What will we treat here, and where shall we start? The earth is home to a bewildering variety of life-forms. Sizes span about nine powers of ten, from 0.1 micron, as in a virus, to 100 meters, as in a large whale. Hundreds of years ago neither the relevant tools nor knowledge existed to usefully seek answers to many fundamental questions: What, exactly, is life, and when and how did it arise? Where is it going? Because of the amazing variety and complexity of life, scientists of yesteryear adopted a relatively modest goal: develop a classification system for life to try to bring more order to the confusing panoply of the earth's organisms. It wasn't until about the mid-nineteenth century that biologists began to tackle truly fundamental questions. Although many answers still elude us, remarkable progress has been made.

In contrast, for example, to the establishment of the paradigm of plate tectonics, research in biology that led to our current model was mostly done by single or a few individuals or in some cases small groups. At present, however, many individual biology research groups are quite substantial in size yet still lag far behind kindred groups in, say, experimental particle physics.

This story of the search for a fundamental understanding of the biological world has so many threads that some (subjective) selection is necessary. The path I chose is based on heredity and evolution and will, I hope, be reasonably clear. There is, though, a problem with my originally stated philosophy: avoidance of jargon; I found it more difficult in this part of the book than in the first two parts. But I intend to clearly define each obscure or technical term at its time of first usage. You, the reader, may judge the extent to which I have succeeded—or, alas, failed.

The evidence for evolution seems overwhelming. Fossil finds show that creatures were extinguished throughout well more than the last half-billion years. Further, species exist now that apparently did not always exist: no fossils of them have ever been found from the distant past. Of course, here that old saw applies—namely, absence of evidence is not always evidence of absence. In the face of such evidence, or lack thereof, who could possibly doubt that some sort of biological evolution has taken place over the history of the earth? (Next funny question!) As in Alfred Wegener's first serious proposal of continental drift, the motor was missing: What *causes* biological evolution? A large part of our story will be devoted to the uncovering of this motor in the context of biology.

We start with the transformation of biology from a science based primarily on classification to one based on observations, processes, and experiments, along with understanding; it is an amazing story. Three major contributions to this revolution were made by four separate scientists in a little over a decade, near the beginning of the second half of the nineteenth century. The stage was thus set for understanding heredity: why elephants beget only elephants, mice only mice, and, thankfully, never any mixed relations.

Charles Darwin and Alfred Russel Wallace

Charles Darwin's background was economically comfortable. His father was well to do and wanted Charles to be a physician. Charles started medical studies but didn't like them, preferring to study and train in botany and geology. In 1831 at age twenty-two, Charles volunteered to be the resident naturalist on a circumnavigation of the globe on the good ship *Beagle*; this trip was expected to last about two years but ended up lasting five, with much of the time spent in the vicinity of South America. Charles was an acute observer of life-forms of all sorts that he encountered, as well as of geological formations. He kept very detailed notes of his observations and thought deeply about them. At that time, people in this field of work were usually either observers or thinkers. Darwin was unusual; he was both.

From his observations, Darwin began to form his views on evolution. The facts he gathered seemed to him inconsistent with a creationist explanation, but entirely consistent with an evolutionary interpretation. He then hit upon the main idea of evolution on which his fame largely lies: the (macroscopic) motor, the principle of natural selection. What is this principle that nearly everyone mouths but by no means everyone understands? In brief, those organisms with traits that enable them to better survive and *reproduce* pass these traits on to offspring so that these favorable traits spread through the population of future generations of these organisms. In that way, this set of organisms slowly evolves and better survives. This principle is often expressed in shorthand as "survival of the fittest," which seems in practice to be no more than a tautology: it would seem that, almost by definition, those who survive are the fittest, unless luck plays the deciding role, even altering statistical likelihood. However, there really is no tautology. The crucial fact about natural selection is that offspring are not perfect copies of their ancestors, differing in particular in their increased ability to reproduce.

From these and other facts that he gathered, Darwin argued strongly against evolution by major steps (saltation, a sudden and large change from one generation to the next) and in favor of small changes, virtually imperceptible from one generation to the next. For example, he used arguments based on the breeding of fauna for domestication where one observes very gradual changes from one generation to the next. The idea of saltation in biology still has some firm proponents, and may in fact be true in certain cases, but is now, at best, still controversial.

Darwin returned from the voyage of the *Beagle* in 1837, and he spent the next two *decades* buttressing and polishing his ideas, trying to anticipate all possible objections and to answer them convincingly. In addition, he had to deal with a number of serious personal problems—health crises of his own and of his children, two of whom died young. While Darwin was still thus engaged, a letter posted early in 1858 by Alfred Russel Wallace from what is now Indonesia arrived in June at Darwin's home. It proposed a theory of natural selection and asked that it be published if Darwin thought it worthy. Darwin apparently showed the letter to three people—Charles Lyell, Joseph Dalton Hooker, and John Joseph Bennett—whom he knew were in prominent positions in the Linnean—or Linnaean—Society. This society, founded in London in 1788 for discussion of advancement in the life sciences, was named after Carl Linnaeus, a famous eighteenth-century Swedish naturalist whose collection of fauna and flora was bought in 1783 by the founder of the society, the Englishman James Edward Smith. Linnaeus's fame stemmed primarily from the biological classification system that he developed.

At least one of the three was aware that Darwin had been working on essentially an identical theory for about twenty years. These people at the society felt that it would not be just to publish Wallace's letter without giving Darwin the opportunity for simultaneous publication. In a masterful, Solomonesque decision, Lyell, Hooker, and Bennett wrote, "We feel it desirable . . . that views founded on a wide deduction from the facts . . . should together be laid before the public." So, at the next meeting of the society, Wallace's letter and a note from Darwin on natural selection were both read. Not much of a public stir was created. Both communications were published that summer in the society's journal, also not creating much public interest. But Darwin was clearly on notice: he had better publish the details of his theory quickly. He worked feverishly over the next year or so to produce *On the Origin of Species,* about a five-hundred-page book, which he considered merely an abstract of his planned more complete version. His publisher nixed this abstract label. The book, by and large, gives no credit to precursors, has no references, and so states. The reason was simple: no time. Darwin rushed into print, as noted, to protect his priority, which desire apparently overcame his competing desire to develop the book at much greater length. Despite his rush, the final words in *On the Origin of Species* are rather poetic: "There is grandeur in this view of life, with its several powers, having been originally breathed into a few forms or into one; and that, whilst this planet has gone cycling on according to the fixed law of gravity, from so simple a beginning endless forms most beautiful and most wonderful have been, and are being, evolved."

Before discussing some of the science behind, and the reaction to, Darwin's book, let us briefly treat the relations between Darwin and Wallace. Given this rather awkward aspect to their relationship, how did they get along? They were apparently very respectful of each other, with Wallace deliberately staying in the background, for whatever reason(s) — none overly clear to me! Possible reasons include Wallace's deference to the greater breadth of Darwin's work, Darwin's being fourteen years his senior, Darwin's greater station in life (he was rich; Wallace was rather poor, certainly by comparison), and Wallace's relatively shy personality.

Whatever . . . . In later years Darwin did use his (considerable) political influence to intercede for Wallace in helping to get him a pension from the government. Even later, they had a falling out over Wallace's backtrack from evolution in regard to the human mind. Wallace somehow gained belief in a supernatural origin for the human intellect, a belief that Darwin found abhorrent. Wallace also backtracked on the origin of life and of consciousness, as both being due to God's intervention as for the human mind.

Back to science. Most everyone has heard of Thomas Malthus and his early 1800s prediction of catastrophe, unless the inexorability of exponential population growth was otherwise checked. It is, in fact, easy to show how reproduction of any species would be expected to soon overwhelm the earth unless otherwise checked. An approximate doubling of the population, for example, with each generation quickly leads to overwhelming numbers of individuals. Nature has many means to provide such checks, for examples pestilences and decreases in fecundity accompanying decreases in per capita resources. Both Darwin and Wallace were familiar with Malthus's work, and each discussed a representative number of such checks in detail, for example, starvation when resources were overtaxed. It has been said that Malthus's proposition, more than any other single argument, was the basis for natural selection. Why did Malthus's proposition have such a profound effect, independently, on Darwin's and Wallace's formulations of natural selection? Darwin, for one, thought after pondering Malthus's argument that his proposition made members of a species competitive and that those who were better prepared to survive and reproduce in this competition would pass on to future generations this competitive edge, thus yielding natural selection as the (macroscopic) motor for evolution.

A quotation from chapter 3 of Darwin's book succinctly displays his ideas on natural selection: "The struggle almost invariably will be most severe between the individuals of the same species, for they frequent the same districts, require the same food, and are exposed to the same dangers."

Darwin noted the great variability of individuals of any given species. What, he wondered, did nature do with all of this variability? Perhaps it uses this variability to enable the species to survive over many generations even though, say, the environment changes a lot. Genetic information passed from one generation to the next is not at all random but is a predictable property of genetic material, whatever this material might be. Natural selection—survival of the fittest—is a direction-giving force. In a very crude, macroscopic, sense, it was the previously missing motor. Darwin fleshed out the idea of natural selection for a wide variety of species and environmental circumstances, as did Wallace in his independent work. It was clear to them that each species' survival depended, not on accident, but on genetic gifts in the population of that species. (All of the individual organisms of a given species alive at a given time constitute the population of that species at that time.)

Only some organisms in each population reproduce, with the number varying with species and circumstance. But success in having progeny is the key; mere survival of an individual is useless in propagating the species, unless that individual, somehow, helps other individuals of the species to reproduce. The motor of natural selection, even though how it operated in heredity was not then understood, dispensed of the age-old issue of teleology, an approach that goes back apparently at least as far as our old friend Aristotle. Crudely speaking, teleology is the explanation of a natural phenomenon in terms of its purpose. For example, we have feet in order to be able to walk. Put another way, teleology is the concept that a characteristic (for example, my foot) exists because it serves a function. Such explanations are in modern times not usually considered to be useful or relevant.

It is certainly true that Darwin was rather vague and somewhat confused about the origin of genetic variation and the detailed process of heredity. He was, though, well aware at least since the beginning of the 1840s of the need for some special mechanisms to govern heredity. He finally created an apparently relevant theory in the middle 1860s; he called it pangenesis, in which "pan" means "whole." This theory was complex and lacked detailed mechanisms; it postulated that each cell in an organism would, in effect, excrete tiny hereditary particles, which he called gemmules, and that these (somehow) aggregated in the reproductive organs. The idea that cells throughout the body needed to transmit information to the reproductive organs likely had its origin in the work of French naturalist Jean-Baptiste Lamarck, who had postulated in the early 1800s that characteristics acquired during an individual's life can be inherited—that is, passed from one generation to the next. This postulate proved wrong. But the point here is not to criticize Darwin (or Lamarck); rather, it is to illustrate how difficult it was in the mid-nineteenth century to gain any understanding of heredity. Now, of course, the mechanism of heredity—that is, its microscopic motor, DNA—is known to every schoolchild who has taken a first course in biology. These insights, however, only came about a century after Darwin's and Wallace's development of the theory of natural selection and were hard-won, as we shall see.

Returning to 1859, we ask: What was the general reaction, from scientists and from the general population, to *On the Origin of Species?* From the modern perspective, it may be fair to say that it hit like a hydrogen bomb. Apparently, most of the elite were very vociferous in their objections; lay-people at first apparently took little notice. "Outstanding" naturalists had been opposing evolution for (at least) the *preceding* three decades, but not very noisily. Why did they make such a big fuss now? My guess is that pre-decessors who had been pushing evolution did not create much of an op-positional stir to this promotion because:

1. Their words/ideas were clearly only vague speculations;
2. They were generally uninformed, unaware, and uncritical about complications; and
3. They inserted their arguments into treatises dedicated primar-ily to other subjects.

By contrast, *Origin* was a well-organized, and well-buttressed, account, based on data collected over a large fraction of the world, presented in one book rivaling the Bible in length, if somewhat different in content. Darwin, as already noted, took great pains to think of all possible objections to his arguments and to provide answers as cogently as he could. For all of these reasons, Darwin was taken seriously. Interestingly, Darwin had no intention of challenging religion in *Origin*. Indeed, he was apparently very careful in his wording *not* to offend those with religious beliefs. I should add that I think Darwin and Wallace were the first evolutionists not to seek refuge in extraobservational bases (such as "God did it"), except for Wallace's later retreat, for example, on the origin of the human mind. Natural selection was simplicity itself.

What about the opponents of evolution? What were their arguments? They very often noted that, for example, breeding of dogs always begat dogs—never an organism of a different species. People could not and did not deny that breeding produces changes, but they drew the line on *species* changes.

First, what *is* a species? The answer is often far from obvious and is often controversial. Here we remark only: the common definition, that two crea-tures of opposite sex belong to the same species if they can successfully mate with one another and produce successfully mating progeny, is, for example, not of much use for nonsexual reproduction and very hard to demonstrate for *extinct* species. Nature, in the form of biology, is incredibly complicated; thus, finding a definition useful over the entire range of organisms may well be a chimerical pursuit.

In any event, a major, very thoughtful critique of evolution was penned by Fleeming Jenkin in 1867. He pointed out that there is no evidence for a change in species, and there is a limit to the extent of change that breeding can bring about—for example, horses can be bred to run only so fast. He also noted that breeding of outliers, those with extraordinary characteristics, often leads to offspring with more ordinary characteristics, a phenomenon sometimes called "regression to the (prior) mean." Jenkins presents his arguments at great length, concluding, "if the arguments of this essay be admitted, Darwin's theory of the origin of species is not only without sufficient support from evidence, but is proved false by a cumulative proof."

What were the responses to these criticisms by Darwin and his strong supporter, the widely respected contemporary intellectual Thomas Huxley, first mentioned in chapter 12? The basic problem was that Darwin did not have a correct model of heredity; in fact, he had only a hazy model, as noted above. Even Huxley did not have a satisfactory answer to the above critique. A key problem was that these objections were based on a tacit understanding of heredity as a continuous, for example, mixing, process. The correct process, a discrete one, digital rather than analog, involving mutations on a molecular level, was unknown to, and unsuspected by, anyone in that era. This (new) fact allows avoidance of "regression to the mean."

Other critics used arguments such as: "Darwin proposes a mechanical, soulless, universe." Always beware of such nonsubstantive-type critiques; they usually signify that the person behind them has no valid argument.

Gregor Mendel

We now leave Darwin, the observer and thinker extraordinaire, and Wallace, and move on to their contemporary Gregor Mendel, the biological experimentalist extraordinaire. Mendel introduced controlled experiments into biology—a major advance. He used this approach to hunt for rules governing the breeding of plants. Many historians viewed Mendel as lucky to have discovered laws of heredity. I think that assessment does Mendel an injustice. The organization of his experiments and his clear presentation of the results may be due not only to his logical mind but to his training as both a physicist and a teacher. That the resultant rules disclosed by his experiments were simple can be considered fortuitous, but that outcome shouldn't, in my view, be tainted by the pejorative implication that Mendel was lucky.

Virtually every educated person has heard of Mendel's experiments with pea plants. Not so many are familiar with the details, which are contained in his magnum opus, published in 1866. In total his experiments involved about ten thousand plants. His paper describes the details of his breeding of these plants in about fifteen pages, and he leaves some possibly important issues untreated or treated only in part. His basic assumption was that the "mothers" and the "fathers" contributed equally to the inheritance of characteristics by the next generation of plants. In the bulk of the paper he showed that some characteristics of pea plants seemed consistent with this assumption and followed simple rules of inheritance. Near the end of his paper, Mendel noted that different characteristics of the pea plant from those he had discussed were not so simply treated and, indeed, he had no idea how to treat them—that is, how to fit them into rules like the simple ones he had already found.

Mendel's experiments in plant breeding extended for about eight years, between 1857 and 1865. How did Mendel choose the type of plants to use for these experiments? He set out the following three criteria: plants must possess clear, easily differentiated characteristics; breeding plants must be well protected during reproduction from foreign influences; and bred plants must show no strong change in fertility in successive generations.

He chose *Pisum sativum* (the garden pea), which satisfies these criteria, and chose seven of their characteristics to study: form of ripe seed; color of seed albumen; color of seed coat; form of ripe pod; color of unripe pod; position of flowers; and length of stem.

Bred sets of pea plants had to be reliable, and easily distinguishable, carriers of one of two distinctive characteristics, for example smooth seeds or wrinkled seeds, and long (1.8–2.1-meter) stems or short (0.3–0.5-meter) stems. To ensure that the plants were pure carriers of only one of the two distinctive, contrasting characteristics, he bred those exhibiting the one characteristic over and over again, selecting at each stage for future breeding only the pea plants with the desired characteristic. Then, when he was sure that he had a large number of plants isolated with one characteristic and a large number of plants isolated with the other of the two contrasting characteristics, Mendel bred the first set of plants with the second set, one from one set with one from the other, producing many offspring.

Mendel's results were striking. All of this first generation of offspring from the mixing of the plants, one of each mixed pair being pure for one of the two contrasting characteristics, exhibited one and only one of the two possible characteristics of any of the seven traits, for example, all long stems and smooth seeds. That is, one of each pair of contrasting traits proved dominant, and the other, by definition, was recessive—that is, it did not appear in any of these first-generation bred plants.

When only these first-generation offspring were bred among themselves, the next generation of plants showed a clear difference in the appearance of the characteristics of any trait: about three-quarters of the plants exhibited the dominant trait, and about one-quarter exhibited the corresponding recessive trait.

These results represented a major advance over those of his predecessors. Instead of studying many different generations of many different plants in a, shall I say, helter-skelter manner, Mendel did one series of well-planned, careful, connected experiments extending over those eight years. He thereby discovered simple rules of heredity missed by all of his predecessors. (In his paper, Mendel included proper credit to his predecessors but omitted critiques that he most likely, I think, had of their experimental procedures.)

What effect did Mendel's results have on the scientific community? He presented his results at a scientific meeting and subsequently published them. They then disappeared from sight like a stone dropped in midocean. What happened? Many historians have placed the blame on a contemporary, Carl Wilhelm von Nägeli, who wrote the major book on relevant evolutionary matters at the time and didn't even mention Mendel's results.

Simon Mawer, a British novelist, born in 1948, who had been a biology teacher most of his life, probably put it best: "We can forgive von Nägeli for being obtuse and supercilious. We can forgive him for being ignorant, a scientist of his time who did not really have the equipment to understand the significance of what Mendel had done despite the fact that he [von Nägeli] speculated extensively about inheritance. But omitting an account [or even mention] of Mendel's work from his [von Nägeli's] book is, perhaps, unforgivable." To be fair, though, I must add that it is possible that von Nägeli did not appreciate Mendel's work, perhaps because Mendel himself apparently considered his results to be relevant to hybridization (mating of two organisms with different characteristics), rather than to heredity, per se.

The Resurrection of Mendel's Work

Mendel's work was resurrected apparently independently and almost simultaneously by three botanists from three countries: Hugo Marie de Vries (1845–1935; Netherlands), Carl Erich Correns (1864–1933; Germany—a student of von Nägeli!), and Erich Tschermak von Seysenegg (1871–1962; Austria)—near the start of the twentieth century, over three decades later. Of these three, apparently only Correns had appreciated the significance of this work. A fourth who also resurrected Mendel's work, I think a bit later, was an American, William Jasper Spillman.

When William Bateson, who coined the term "genetics" for the study of heredity, learned of Mendel's work as a result of this rediscovery, he became outraged at the injustice to Mendel, who, he felt, should have received great acclaim for his findings. He changed the emphasis of his own work on heredity, based on Mendel's findings, which he—Bateson—appreciated in full for its implications for rules of heredity. Some years later still, in 1909, Bateson wrote a rather widely distributed book on Mendel's work. This book led to Mendel's penetrating, and remaining in, the consciousness of scientists and of all students of genetics, including those at the high-school level. PR-like work had amazing effects even then.

Bateson's colleague Reginald Punnett invented Punnett squares, which succinctly summarized Mendel's main results (fig. 16.1). In the example

Punnett Square

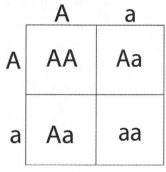

Figure 16.1. A simple Punnett square. Courtesy of David Shapiro.

shown here, the top row (above the two-by-two matrix) indicates that for the trait "A," one of the pair being bred (or mated) has, in modern parlance, one dominant, A, and one recessive, a, gene. The first column (to the left of the two-by-two matrix) indicates the same for the other of the pair being bred. The entries in the two-by-two matrix show the proportion of the offspring that will have each complement of genes. Thus, as shown, one-quarter of the offspring will have *AA* genes, one-quarter will have *aa* genes, and one-half will be *Aa* (or, equivalently, *aA*). These pictorial representations of Mendel's hereditary rules, now called laws, are still widely used well over a century later.

When more sophisticated statisticians in the first third of the twentieth century started pouring over Mendel's results, criticisms arose. Claims were made that Mendel's results were highly improbable and, in fact, unbelievable. This point of view was expressed by R. A. Fisher, a towering figure in the field of statistics in his or any day. Mendel's data matched his laws more closely than probability theory indicated was at all reasonable. Statisticians thrusted and parried over Mendel's results for nearly a century, though this controversy seems to have recently died down, or at least I have not seen any articles appear on this subject in the past few years (not that I've done a thorough search . . .). The nub of the issue, as often in these matters, relates to critics' assumptions and their bases. The details here are arcane. One aspect, however, is clear: no one accused Mendel of fraud; quite the contrary. Possible explanations offered included unconscious bias (placement of doubtful plants aside, thus favoring his hypothesis); disregarding data thought to be contaminated that were merely statistical fluctuations; and his assistant did it, for which there is no evidence.

In the next chapter we move on to the search for the molecular basis for heredity, with its culmination reserved for the following chapter.

SEVENTEEN

The Molecular Basis of Heredity

The theme of this chapter is the search for a molecular basis for heredity. As with many, but not all, of our detective stories, we now know
the answer (or at least its broad outline). However, it helps one's perspective
and understanding—my rationalization—to follow the trail that led to the
pot of gold at the rainbow's end.

We tell this story by addressing two (compound) questions: What is the
key molecule (or molecules) that determines heredity, and how do we know?
How does it (or they) work, and how do we know? As is our wont, we pursue
the details via the main contributors, in historical order. In this chapter, we
treat the principal first five, covering from the 1860s to the 1940s. These five
in my subjective view are:

> Friedrich Miescher (chemist)
> Phoebus Levene (chemist)
> Frederick Griffith (medical officer)
> Oswald Avery (biologist-chemist)
> Erwin Chargaff (chemist)

To set them in their rough historical places, I note that the first was born in
1844 and the last died in 2002.

The Discovery of DNA

As almost every schoolchild now knows, the principal molecule of heredity goes by its initials, DNA. Few schoolchildren—or adults—know that its proper chemical name is deoxyribonucleic acid. Virtually no one knows its atomic structure. And, I will wager, even fewer have heard of its discoverer, not that this person's name is needed to understand the role of DNA. But it is a most famous molecule, and its discoverer is known to almost no one. Let us correct that anonymity now: the discoverer's name is Friedrich Miescher, born in 1844, another loner, from Switzerland. His family contained several distinguished scientists. In fact, Miescher went into chemistry because of his uncle's conviction that the last remaining open questions on human tissue development could be solved via chemistry.

Miescher first went to work in 1867 in Germany in the laboratory of Ernst Felix Hoppe-Seyler, who had done fundamental work on hemoglobin. Miescher chose to work on the chemical composition of cells, another step toward accepting his uncle's advice. Hoppe-Seyler suggested that he study lymphocytes (white blood cells); Miescher set out to study the nuclei of such cells.

To pursue his studies, Miescher invented many of his own protocols, mostly via the technique known fondly as trial and error. He had daily access to used bandages from a nearby clinic, thus providing himself with a plentiful supply of white blood cells. He separated these cells by soaking the bandages in a (nine-to-one water to sodium sulfate) solution. After several days, the desired cells settled to the bottom of the container. Miescher identified five proteinlike entities in these cells via their solubility properties—the degree to which they dissolved in various liquids. He also found one other substance, unlike any known protein. Serendipity struck again. This new substance came from the nuclei of cells (see below) and had unexpected properties, which Miescher exhibited by using chemists' black magic of the times: he found that it precipitated—went to the bottom of its container—when he immersed it in an acidic solution, and then redissolved when he made the solution more basic (the opposite of acidic). Finally, he discovered that prolonged—for the order of a day—exposure of cells to very dilute hydrochloric acid produced a residue in the test tube resembling the nuclei of cells. Little was then known about the cell nucleus, other than its appearance, despite its having been discovered in 1719 by Antonie van Leeuwenhoek, the inventor of a useful version of the microscope, and its description having been given for orchid cells by Robert Brown, of Brownian motion fame, in 1831.

Miescher tried to stain this new nucleus material with iodine, which was then thought to cause proteins to turn yellow. No yellow color resulted. He concluded that this new substance was not a protein. Further study implied to Miescher his need to increase the purity of this mystery substance. To meet this need, he developed complicated protocols involving, for example, warm alcohol, ether, and acid from a pig's stomach, which contains the enzyme pepsin. (An enzyme, by the way, is a molecule that catalyzes—speeds up—a specific chemical reaction, often the breakdown of other molecules; the enzyme itself is unaffected by the reaction.)

Miescher's impressive persistence paid off: he found that the extracted nuclear material sank to the bottom of his test tube in fine, white granules. What was the elemental composition of this new precipitate? With his final call on chemistry, Miescher used the chemical methods of his day to isolate the percentage of each element in this precipitate. As a precaution to ensure that his procedure was sound, he weighed the reaction products—the elements he had successfully separated—and compared the total of their weights with the weight of the total of the materials with which he began, to check on whether he had lost any of the original materials or had introduced any (significant) contaminants. He was satisfied with the reliability of his results but nonetheless startled by his findings. In addition to carbon, oxygen, hydrogen, and nitrogen—all expected elements—there was a large amount of an unexpected element: phosphorus, almost nonexistent in then known organic materials. Further, this precipitate was composed of a new molecule, which was very big: it would not diffuse through parchment, and therefore he believed that it had a large molecular weight, at least 500. (Molecular weight is the weight of a molecule usually expressed as an integer, which represents the total number of neutrons and protons in a molecule's atoms. We are ignoring here such small differences as between a proton's and a neutron's mass.) By way of comparison, a molecule of ordinary water, H_2O, has a molecular weight of 18.

Miescher realized that he had discovered a fundamentally new type of molecule. He called it nuclein, based on its origin in the nucleus. Thus came about the serendipitous discovery of what is now called DNA. It is intriguing to me to see the large role played by serendipity in the advance of science. It even happened once in my own career: my discovery, with my students and colleagues, of radio sources in the cosmos that appeared to be moving at speeds greater than that of light. But that is another story which, by the way, includes a rather magnificent demonstration of the validity of Einstein's special theory of relativity rather than a refutation of it.

What were some contemporary speculations about the nucleus? In 1866, Ernst Haeckel, a famous German biologist, presciently suggested that the cell nucleus might be responsible for the transmission of hereditary traits. Miescher was confident that his new molecule, nuclein, would prove of equal stature to proteins in its role in the cell; alas, he scrapped his initial idea that nuclein was solely responsible for the diversity of species, although he did speculate that nuclein might provide sufficient variations to explain diversity, in analogy with words and letters, as we'll discuss later.

What about publication of his startling new discovery? Miescher had discovered a molecule wholly unlike any other known molecule, or type of molecule, then recognized in the world of biology. Publication should have been a cinch, right? Wrong (or I wouldn't have raised this question!). Hoppe-Seyler demanded two prerequisites to allow the submission for publication: first, Miescher had to broaden the base of applicability of his discovery by showing the presence of nuclein in cells of another organism; and second, Hoppe-Seyler himself had to reproduce Miescher's results. Miescher struggled but eventually showed the presence of nuclein in the cells of salmon, and Hoppe-Seyler did succeed in reproducing Miescher's results. All in all, it took two years for Miescher to obtain permission to publish. However, his startling discovery was apparently not pursued by others in his era, yet without a clear villain to be blamed, as in Mendel's case.

Although not of robust health, Miescher still drove himself very hard research-wise and succumbed, apparently to tuberculosis, in 1895 at the age of only fifty-one. He had not been much of a self-promoter. His fundamental contribution dropped from sight, despite his uncle's publication of Miescher's collected works with the introduction: "The appreciation of Miescher and his works will not diminish with time, instead it will grow, and the facts he has found and the ideas he has postulated are seeds which will bear fruit in the future." That future was more than eighty years in the making.

Why isn't Miescher, the discoverer of what is now called DNA, a household name like Darwin, Mendel, James Watson, and Francis Crick? I am aware of no good answer. Perhaps the gap of over eighty years between his discovery and the understanding of its importance was just too long. Or, perhaps, there was just no one promoting his contribution, such as Bateson, who devoted so much of his time to promoting Mendel. I just do not know. Some people think it is because he did not establish its role in heredity; I think that argument is asking too much and should not be used to detract from his key discovery.

Progress and Then a Wrong Path

Next, let us skip ahead a few decades to the early twentieth century and our next principal character, Phoebus Levene. He was a superb organic chemist who was born in Lithuania, grew up in Saint Petersburg, and then emigrated to New York City in 1893, forced with his family from Russia by anti-Semitic pogroms. In New York, he was employed by the Rockefeller Institute of Medical Research. Phoebus Levene made important contributions—both positive and negative!—to our understanding of Miescher's nuclein. (Hereafter, I will use only the modern designation, DNA.)

Miescher had determined the relative amounts of five elements in the DNA molecule but not its chemical structure. Enter Levene, and others, with more modern chemistry techniques at their disposal. Levene characterized the chemical building blocks of DNA, four entities called bases: adenine, guanine, cytosine, and thiamine. A two-dimensional picture of each of these four bases shows their constituent elements and relative placements, along with their classifications as either of two types of molecules: purines or

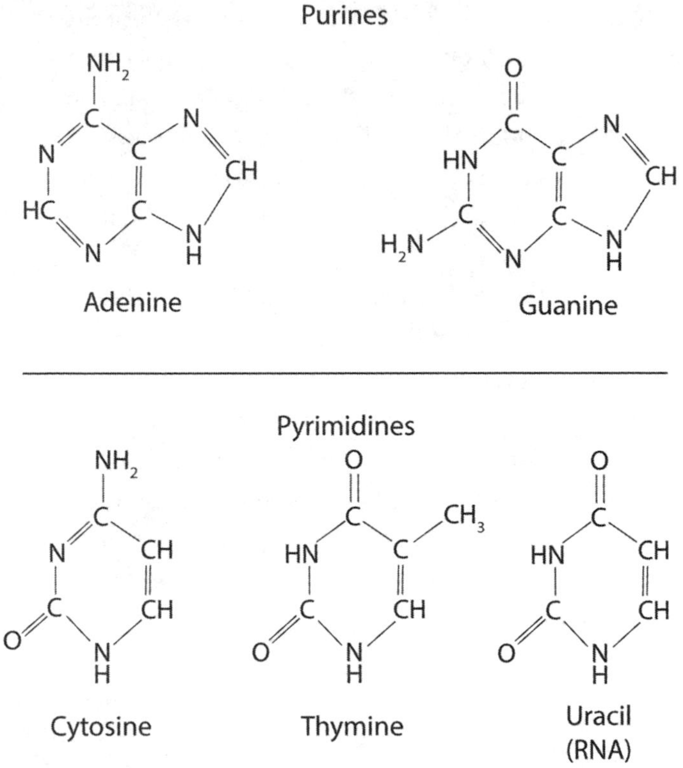

Purines

NH₂

Adenine

Guanine

Pyrimidines

NH₂

CH₃

Cytosine   Thymine   Uracil
(RNA)

Figure 17.1. The chemical element structure of DNA bases: purines and pyrimidines. Courtesy of David Shapiro.

pyrimidines (fig. 17.1). Purines are characterized, roughly speaking, by having a pentagon (five-sided figure) attached to a hexagon (six-sided figure), and pyrimidines by only the hexagonal component. Note that there is a fifth base, uracil; it is not contained in DNA but is in RNA (ribonucleic acid). We will discuss uracil's biological role later.

Inspection of this illustration discloses the differences in chemical composition, and in structure, between similar types of molecules, for example the various pyrimidines. Thus, thymine and uracil differ only in one respect: where uracil has hydrogen attached to a carbon atom, thymine has CH_3. What these differences imply for the chemical properties of the bases is the important issue (but is beyond the scope of this book). DNA also contains the sugar, deoxyribose, and the phosphate group. (Deoxyribose differs from ribose by not having an oxygen atom attached to one of its hydrogen atoms; the key question, not addressed here, is the effect of this missing atom on the properties of deoxyribose and its biological importance.)

In addition to this major advance in determining the basic arrangements of the elemental constituents of DNA, Levene also claimed that its components were connected in a particular order—phosphate-sugar-base—to form units. He labeled each such unit "one nucleotide" and concluded that the DNA molecule was a string of such nucleotide units linked together through their phosphate groups, which formed the backbone of the DNA molecule. Levene went further, apparently formulating the concept called the tetranucleotide hypothesis, in about 1909. This hypothesis proposed that DNA was made up of *equal amounts* of adenine, guanine, cytosine, and thiamine, with four of them—one of each type of base—constituting each tetranucleotide. This hypothesis was apparently not seriously challenged for over thirty years. It led, unfortunately, to the conclusion that DNA could not possibly be the carrier of genetic information. Why not? For the actual hypothesized structure, there just wasn't anywhere near enough variety possible for DNA to be the carrier of as much information as would be required for producing organisms, especially for those as complicated as humans.

If not DNA, what other molecule(s) had enough varieties to hold the secrets of heredity? Biologists then thought that proteins must be the basis of heredity because of the huge variety of possible protein molecules. What is a protein molecule made of? Amino acids. What constitutes an amino acid? The chemical structure of a typical amino acid is shown in fig. 17.2. About five hundred amino acids are known. But for some unknown reason(s), only a specific set of twenty are mainly used in biology. In some biological systems, two more amino acids are found naturally. Biology is *very* complicated, and there is a *lot* we don't understand, probably even in cases when we think we do understand.

Proteins are composed solely from sequences of amino acids and form rather bizarre three-dimensional shapes depending, in part, upon their environment. In a sense, amino acids in biological systems are the letters of a (twenty-letter) alphabet and proteins the words. However, for a word in a language, only the linear order of the letters is relevant; for a protein, the

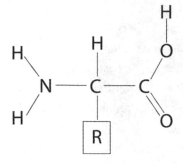

Figure 17.2. A generic amino acid's chemical (element) structure. The "R" symbolizes the various elements that characterize the different amino acids. For example, the simplest amino acid is glycine, for which "R" represents a single hydrogen atom. Glycine has recently been discovered on a comet via a European space mission. Courtesy of David Shapiro.

specific three-dimensional structure of the protein molecule is also of great importance. Even were proteins to each be twenty amino acids in length—most are far, far longer—there would be 20^{20} chemically different possible proteins, a staggeringly huge number. A human cell can contain upwards of about one hundred thousand different kinds of protein molecules, with the total number of protein molecules—same and different—in a cell being as much as of the order of one hundred times as many! Not that anyone has actually counted: these are just crude estimates but may be in the right ball park. The number of amino acids in a single human protein molecule apparently varies from the order of fifty to about thirty-four thousand.

The tetranucleotide hypothesis and the hypnotic effect of the enormous variety of possible proteins seemed to set biologists on a wrong track for several decades in their search to understand heredity on a molecular level. Nonetheless, one must give Levene great credit for his important contributions to the understanding of the organization of the elements in the chemical structure of DNA.

The Transforming Principle

The question still remained: What exactly is the genetic, or hereditary, material? If proteins, how, exactly, do they determine the characteristics of organisms, and how do they pass this information from one generation to the next? New progress toward answering these questions began in the late 1920s from an unlikely source, a medical officer of the British Ministry of

Health, Frederick Griffith. His specialty was pneumonia. How you might wonder—if you didn't already know—was expertise in pneumonia relevant to the unraveling of the heredity enigma?

Griffith's main concern was to understand the differences in the virulence of this disease of different pneumococcal types. His goal was to better treat and, perhaps, prevent this disease from spreading or even from infecting humans. Griffith studied varieties of four types of lobar pneumonia and presented his results in 1928 in a report on 278 pneumonia cases in which he especially studied sputum samples—saliva mixed with mucus coughed up from the lungs of the suitably ill. Puzzled by some of his observations, he sought understanding by conducting a large number of experiments under a wide variety of conditions and by pretreatments of mice with various strains of pneumococci. Griffith was a serious and careful experimenter.

Griffith noticed that some colonies of pneumococci had rough surfaces, the "R" form; these pneumococci were not virulent: they never killed their hosts. Other colonies had smooth surfaces, the "S" form; these were generally virulent. That is, when mice were injected with the S form of pneumococci, they generally died. The big surprise was the following: R forms were convertible to S forms when mixed with killed (via heating) S forms, by injecting both into the body of a live mouse. The question: What was the transforming principle? What, exactly, allowed the heated, thus killed, S forms to convert the R forms to S forms in the body of a mouse? Griffith brought about these conversions by a variety of protocols involving inoculations of mice in different parts of their bodies. Was this behavior of the pneumococcal bacteria inside of a mouse medical black magic, or was there some fundamental property that could and should be pinned down?

Griffith's results soon became widely known and were the subject of much comment and inference but little or no new experimentation, mostly near repetition of his original experiments by likely incredulous biologists. They nonetheless confirmed Griffith's results in the early 1930s, including inducing this strange transformation in vitro—that is, outside of a living organism in a test tube (or equivalent; "vitro" for glass versus "vivo" for living organism). In addition, by the middle 1930s, the transformation of the pneumococcal bacteria from a benign to a virulent form had been effected in a second organism, a rabbit. There was by then absolutely no doubt that this transforming effect was real and needed to be understood.

What were some early views? Griffith suggested (1928) that some specific protein might serve as pabulum (food) to enable the R form to transform. Theodosius Dobzhansky (1941), a world-famous geneticist, for example, for his work on the synthesis of evolutionary biology and genetics, stated, "If this transformation is described as a genetic mutation—and it is difficult to avoid so describing it—we are dealing with authentic cases of induction of specific mutations by specific treatments." Not exactly a precise theory, though aiming rather obliquely at the mechanics of heredity. Other explanations were provided between these two, but none was backed up by experiments, as noted.

The story continues with Oswald Avery, Colin MacLeod, and Maclyn Mc-Carty. This triumvirate, seemingly led by Avery, decided to seek and to study the chemical nature of the substance(s) inducing the transformation—that is, the conversion—of pneumococcal types. Their approach was to isolate, purify, and test the chemical agent(s) responsible for the transformation in vitro, to better control the experiment and its variables. Proteins, carbohydrates, and lipids from the S form could all be removed or destroyed, and yet the remaining component(s) still converted R bacteria. And when Avery and his colleagues took the S form of the bacteria and treated it in a manner reminiscent in general arcana of Miescher's protocol in isolating DNA, they used this isolate of DNA and obtained transformation! When they eliminated proteins or RNA from the R form and heat-killed S-form material, transformation also took place. But when they eliminated DNA from the same material, transformation did not take place. Thus, Avery, MacLeod, and McCarty concluded, by elimination, that the transforming substance must be DNA. This was a major advance, even though the detailed chemical actions of DNA in their study were wholly unknown.

Their experiment was painstaking, with every effort made to avoid contamination. To obtain 25 milligrams (450,000 milligrams is about 1 pound, as noted earlier) of the purified agent capable of transformation, they had used 75 liters (1 liter is about 1 quart) of culture, which thus yielded only a tiny fraction—about three parts in ten million—of the purified substance!

Their protocols (procedures) were rather complicated in detail, as was their discussion in their now classic paper, published in 1944.

Likely because of the novelty of their protocols as well as the complexity of the experimental details, their results were not readily accepted by the relevant scientific community. Critics felt there was still, despite the authors' great care, the possibility of contamination having vitiated their conclusion that the transforming principle was DNA. Avery, MacLeod, and McCarty were themselves very cautious. Their paper concluded with the comment: possible "biological activity is not an inherent property of nucleic acid but [could be] due to minute amounts of some other substance absorbed to it or so intimately associated with it as to escape attention." Scientists are prone to such skepticism, a useful trait when applied in moderation. But one does not readily know in advance where to draw that line separating "moderation" from "excessive."

Let us summarize. The detailed chemical actions of DNA in this study were wholly unknown. Only known was that DNA was the culprit for the conversion of the R to the S form of the pneumococcus. Avery, MacLeod, and McCarty's paper is now considered a landmark in biology because of the procedures used in the experiments and the great care exercised to prevent contamination that could affect the results. This paper seemed to establish with little doubt the relevance of DNA, as opposed to that of proteins, to the secret of heredity. As noted, it was not so universally viewed at the time. The authors' caution plus the skepticism (of others) implied no Nobel Prize. My guess is that their protocols were too complicated, and that too many people were then still enthralled with the possibilities for proteins to be the providers of the hereditary controls, for the authors' work to be widely accepted by the relevant scientific community. But the wall against DNA being the hereditary agent was weakening.

Determining Base Ratios and Demolishing
the Tetranucleotide Hypothesis

During the same year that Avery and colleagues' article was published, a book appeared entitled *What Is Life?* The author was Erwin Schrödinger, one of the two independent inventors, or discoverers—take your choice—of quantum mechanics in the mid-1920s. He noted perceptively that hereditary information could be expressed using a code with a small number of basic parts and still account for life's diversity, in analogy with letters, words, and books. He was doubtless unaware of Miescher's similar idea from about seventy-five years earlier. Schrödinger, like Miescher, turned out to be correct in using this analogy, as we'll see, but neither made any direct contribution to unraveling the story of heredity, except that Miescher discovered DNA and determined the elements that made up its composition.

Schrödinger did, however, influence someone who had a direct effect, our fifth principal contributor, Erwin Chargaff (1905–2002). Having read and been impressed by Schrödinger's book, and having learned about Avery and colleagues' results, Chargaff, then at Columbia University having emigrated from Austria, turned his research program around on the proverbial dime to tackle the chemical attributes of DNA. He considered that lack of specific methods of characterization of DNA and related molecules hindered progress. He thus sought to change that condition and, with his students and colleagues, used new methods to do so. As noted in the Introduction, new technology fuels new science, which in turn often fuels new technology, thus showing a tight coupling between the two.

These new methods were in essence combined by Chargaff and his collaborators into three steps. First, they separated DNA into individual components (mainly the nucleotide bases) by using the new technique of paper chromatography, whose description we skip. Next, they identified the separated bases through the characteristic frequencies of their absorption lines in ultraviolet spectra (frequencies higher than those of blue light). Finally, they estimated the quantity of each base through its relation to the amount of the absorption. The results of Chargaff's and his colleagues' analysis of DNA are simplicity incarnate: in DNA, the number of guanine (G) components is equal to the number of cytosine (C) components (C:G = 1:1), and similarly for adenine (A) and thymine (T) (A:T = 1:1). For the DNA of different species, there are different numbers for the ratios of these two pairs, C&G to A&T, but the same equalities hold for the numbers for each pair. The tetranucleotide hypothesis was thus dead: each of the four different nucleotides was *not* represented equally in DNA in each organism as required by this hypothesis.

From our perspective of now knowing the structure of DNA (see chapter 18), and the vital role played by base pairing, we are presumably faced by an enigma: Why did Chargaff not make the connection between his results and base pairing in the DNA structure? My answer: it was not really obvious, given the lack of knowledge in that day of the structure of DNA. We will, in the next chapter, see the role that this pairing idea played in the unraveling of the structure of DNA by Watson and Crick. In fact, they were aware of Chargaff's results but apparently also did not realize their importance to the DNA structure question at the time Chargaff presented them with his results or for quite some time later—more than a year.

One final gratuitous comment: I suspect that Chargaff's later cynical and sardonic remarks on the issues related to the DNA structure resulted at least in part from his (in my view, understandable) failure to infer the base-pairing idea from his results. He may have also blamed this "obvious" failure for his having been left offstage as awards were given out.

EIGHTEEN

DNA

The Hereditary Molecule and Its Structure

B y the mid-1950s, pretty much everyone had been convinced that we knew the basics of heredity, in particular the structure of the culprit molecule, DNA. Yet even following Avery, MacLeod, and McCarty's work on the molecular basis of the bacterial transformation, and the death of the tetranucleotide hypothesis at the hands of Chargaff, there was still considerable skepticism in the scientific community about DNA's role in heredity versus that of proteins. Confirmation of the molecular material responsible for controlling heredity required another key experiment, one giving an unambiguous answer, to the extent such is possible! We now describe the experiment filling this bill — the Hershey-Chase experiment.

Knowing the chemical elements of proteins (hydrogen, carbon, nitrogen, oxygen, and sulfur) and of DNA (hydrogen, carbon, nitrogen, oxygen, and phosphorus), Alfred Hershey and Martha Chase hit upon a very clever approach to obtain a definitive answer to the question posed. They realized that DNA had phosphorus and proteins (at least some of them) had sulfur, but neither type of the two molecules had both of these elements. Thus, we could distinguish the two, DNA and proteins, by using these two elements to determine which of the two was present in the hereditary material.

To trace these elements, the experimenters made excellent use of discoveries in physics (another example of the unity of science, although I strongly doubt that Hershey and Chase thought of their experiment as using this concept!): a radioactive isotope of phosphorus, ^{32}P, and a radioactive isotope of sulfur, ^{35}S. (Note that the preceding superscript denotes the total number of protons and neutrons in the nucleus of the element whose chemical symbol follows this superscript.)

The next key point was the use of viruses, of which most are composed solely of DNA (or RNA) and proteins, and which depend entirely on other organisms, such as bacteria, for their reproduction. In that era it was well known in the biology community that viruses land on a bacterium, which (unwittingly!) makes duplicate viruses from material injected into the bacterium by the viruses. What Hershey and Chase needed to determine was whether the virus they chose injected its DNA or its proteins (or both) into the bacterium to reproduce. An illustration of a virus, with the location of the DNA indicated, is shown here (fig. 18.1); the rest of the virus is composed of proteins. One usually finds a "global" attack of hordes of these viruses on a single bacterium cell, the tail fibers attaching themselves to the outside of the bacterium.

How did Hershey and Chase get the radioactive materials inside the viruses (many copies of their chosen virus)? They first bred them within

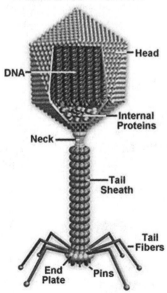

Bacteriophage Structure

Figure 18.1. A virus (a bacteriophage, a bacteria-eating virus), with its major parts labeled. Its total height is about 0.2 millionths of a meter. "Molecular Expressions.com" at Florida State University Research Foundation.

specific bacteria. These bacteria had previously fed for several generations on the relevant radioactive food. After sufficient time, with then standard techniques, the now radioactive viruses were extracted from the bacteria for use in the experiment.

For this use, the viruses were allowed to attach themselves to radioactive-free bacteria. The viruses stayed attached to these bacteria with the virus's tails while the virus material injected through the tails reproduced the virus within the host bacteria. How then do we separate the attached viruses from the bacterium to determine what's inside the bacterium without interference from the viruses? The answer is not obvious! Hershey and Chase conceived of many ways, tried them all, and none worked well. A coworker, Margaret MacDonald, proposed that they use the then and still common kitchen appliance—the (Waring) blender, which my mother had and I loved to use to make and then drink chocolate milkshakes. With this blender, Hershey and Chase successfully separated the viruses attached to the outside of the bacteria from the bacteria.

In a few words: the results from this 1952 experiment were unambiguous; the phosphorus from the DNA in the viruses infected the bacteria cells; the sulfur from the protein in the viruses did not infect the bacteria cells. The case was pretty well closed: DNA was clearly the source of hereditary information, except for the very unlikely possibility that this virus was an exceptional case from the general rule. Here we can consider having demonstrated that a good experiment is worth more than a thousand words!

The Structure of DNA Unraveled

We now proceed to another detective story for the ages: the unraveling of the structure of DNA. A curtailed, and somewhat subjective, cast of principal characters is Francis Crick, Jim Watson, Rosalind Franklin, Maurice Wilkins, and Jerry Donahue.

The main perceived competition, at least by the first two characters, was Linus Pauling, a then world-famous chemist at Caltech, who had recently won additional fame for deducing the single-helix structure of some proteins. A key part of his approach, later adopted by Watson and Crick, was to make Tinkertoy-like models of the proposed structures to see which did, and which didn't, match the data that guided the search.

First let's describe some of the background to the making of X-ray images of DNA, which turned out to be essential for figuring out DNA's structure. In the 1930s, William Astbury of Leeds University in England pioneered the making of such X-ray images of biological molecules. It was far from simple to prepare a particular type of molecule for such studies. One needed many copies of the molecule, together, to be able to obtain an X-ray image. From his X-ray studies of DNA molecules, Astbury deduced that DNA's structure was regular and repeated every 2.7 nanometers (remember that a nanometer is 10^{-9} meters in length) and that the bases (nucleotides) were stacked 0.34 nanometer apart, similar to the spacing of amino acids in proteins. (By point of comparison, I note that an atom's size is of order 0.1 nanometer.) From his data, however, Astbury could not go any further in his attempt to deduce the structure of DNA.

Next came Sven Furberg, a Norwegian working in the middle to late 1940s in Birkbeck College London. He proposed, in his doctoral thesis, based on his X-ray data, that DNA had a helical structure and that the bases were positioned inside the phosphate backbone. He also confirmed Astbury's conclusions on the repeated structure of the bases, but with this repeat cycle related to a full turn of the helix in his—Furberg's—model: eight bases stacked per turn of the helix (8×0.34 nanometer, or about 2.7 nanometers). He also corrected an important incorrect deduction of Astbury: Furberg noted that the sugar of DNA was perpendicular to the nucleotide, not parallel as in Astbury's model. Furberg was finally able to publish his thesis work in late 1952. Why publishing took so long is a story in itself, which we skip.

In an independent project, William Cochran, Crick (the same one!), and Vladimir Vand worked out a mathematical theory of the X-ray images of helices, since there was then renewed interest in the possibility of helical structures being of biological importance. Their paper also appeared in 1952.

I emphasize that this work on X-ray images and on the structure of DNA was going on when it was not yet completely clear that DNA was the carrier of the hereditary information. Publication of the Hershey-Chase experiment then lay in the future.

Our story now shifts back to Cambridge, England, in 1952. At the university, two people met and decided they had common interests. One, Francis Crick, was still a graduate student, although thirty-six years old, and the other was Jim Watson, a second-year postdoctoral fellow from the United States, barely twenty-four years old. Both had not been satisfied with the (different) problems that they had been working on, and, soon after meeting, they decided as kindred souls with complementary knowledge to go for the gold—determine the structure of DNA, which they believed was the molecule controlling heredity. Neither was much of an experimenter, or any at all. They were thinkers who decided to think about models.

Previously, in 1950, Maurice Wilkins, at King's College London, had made an X-ray image of DNA, which led him and Alex Stokes, a theoretician there, to conclude that DNA had a helical structure. They were apparently unaware of Furberg's similar, earlier, proposal, based on his as-yet unpublished work. Wilkins presented this result in the spring of 1951 at a meeting in Naples, Italy, also attended by Watson. Watson apparently considered that this fact would make the full structure of DNA easy to model—not exactly an accurate consideration, as it turned out!

Returning to King's College, we find that in 1951 Rosalind Franklin came to work there on the X-ray study of DNA fibers, as outlined to her in a letter of December 1950 from the director, John Randall, who also assigned a graduate student, Raymond Gosling, to work with her. But Maurice Wilkins, who had been at King's for several years by then, had been led to believe by Randall that Franklin would be working under his general supervision, and he was already studying the X-ray properties of DNA; Randall did not tell Wilkins about the content of his—Randall's—December 1950 letter to Franklin. This lack of communication set a bad tone, and Franklin and Wilkins, when Wilkins returned from his spring's travel, did not have much of a relationship thereafter, illustrating, as we'll see, how poor communication and (consequent) personality clashes can radically change the atmosphere, and even the outcome, of a research program.

Back to the science: Franklin and Gosling noted that there were two forms that DNA could take, depending on the humidity of the environ-

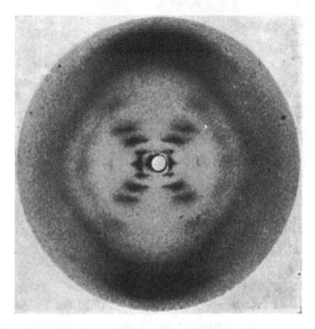

Figure 18.2. The first truly diagnostic X-ray image of DNA, Photo B51, obtained by Rosalind Franklin and Raymond Gosling in 1952. King's College London, College Archives, KCL Department of Biophysics records KDBP/1/1.

ment. When dry, the DNA would assume the so-called A form and when rather wet, the B form, which was longer and thinner, and suffused with water. This realization of two forms could also explain why Astbury's X-ray pictures of DNA were so fuzzy: his DNA may have been a mixture of the two forms. After studying the X-ray patterns of the A form for a while, Franklin, with Gosling, in May 1952, took a remarkably clear and revealing X-ray photo of the B form, which they labeled B51 (the fifty-first photo they took of the B form??) (fig. 18.2). Gosling showed it to Wilkins as part of his graduate student work.

Now back to Watson and Crick. They believed, as noted earlier, that they were in a vitally important race to uncover the structure of DNA. With whom were they racing? They were aware of the world-famous chemist Linus Pauling at Caltech, also mentioned earlier. The other competition of which they were aware was the King's College group(s) of Franklin and Wilkins.

Pauling, for his part, had intended to visit London to attend a conference but was denied a passport because his (leftist) political views had run afoul of the U.S. House Un-American Activities Committee. He then wrote to Wilkins to ask for a copy of this X-ray photo about which he had heard, but Wilkins declined. Without having seen this photo, Pauling came up with a model of DNA involving three intertwined helices, with the backbone of phosphate and sugar on the inside and the bases pointing out. He was aware of the race that he was in: he had published an announcement in the fast-publishing journal *Nature* that he had a new, three-helix model for DNA whose details would soon be published in the slower-publishing journal the *Proceedings of the National Academy of Sciences*, in the United States. It is the only case of which I am aware that someone published an announcement of a forthcoming publication of a scientific paper.

Watson and Crick obtained news in early January 1953 of Pauling's paper from his son, Peter, who was visiting Cambridge and had a copy with him. It was similar to a three-helix model that Watson and Crick had developed less than a year earlier but was shown to be incorrect by Franklin. Then Watson and Crick were enjoined by Lawrence Bragg, their laboratory director, from any further pursuit of a model for the structure of DNA; they were told by Bragg to concentrate instead on their other work. However, upon learning that Pauling was in this race, Bragg rescinded his enjoining order on Watson and Crick; it was then clearly too important a prize to run the risk of Americans capturing it, if the British—even with an American partner—had a reasonable chance.

In any event, Watson and Crick had not lost their interest in the subject. Far from it. Near the end of January 1953, Watson went to London and showed Pauling's paper to Franklin, who immediately realized that it could not be correct, as she had realized for Watson and Crick's very similar model of less than a year earlier. Watson and Franklin then got into some sort of an argument, which apparently Wilkins overheard; he and Watson then talked. During this conversation, Wilkins showed to Watson Franklin and Gosling's Photo B51, without either Franklin's permission or her knowledge. According to Watson's account, his jaw immediately dropped as, by then, his knowledge of crystallography was deep enough for him to realize that this photo indicated that DNA had a *double*-helix structure.

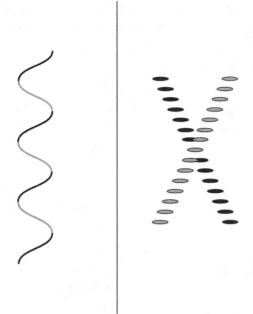

Figure 18.3. The connection between the helical structure of an object, left, and its X-ray photograph, right (see text). Courtesy of David Shapiro.

The "X" formed by the horizontal line segments indicates the helical shape of the object being X-rayed. The angle between the two arms of the "X" gives the "pitch" angle of the helix, essentially related to the number of "turns" of the helix per unit of length along the axis of the helix. The arms of the "X" are perpendicular to the corresponding (main) directions of the (projected) helix (fig. 18.3): the black arm of the "X" on the right being perpendicular to the direction of the black segments of the (projected) helix on the left part of the figure. Hence, the bigger the opening angle (at the top) between the two arms of the "X," the flatter the helix—that is, the fewer the full turns of the helix per unit length along the axis of the helix. Although it is by no means obvious, a missing horizontal line segment in each arm of the "X" in Photo B51 (see fig. 18.2) indicates that there were two helices and which line is missing, in this case the fourth, allows deduction of the spacing between the two helices in the DNA molecule along the direction of the helix.

Why is the center of Photo B51 whited out? Most of the X-rays pass through the sample unobstructed and would dominate the picture, by "spreading" as in photos of that day, overwhelming the information on the structure of the sample. So, the center of the X-ray image is blocked to prevent this spreading; it's analogous to the result of Rutherford's scattering experiment that disclosed the tiny nucleus of the atom: the vast majority of the rays went through the gold foil unscattered.

Watson and Crick relaxed—but only a little bit—when they realized that Pauling had gone down a wrong path with his three-helix model, and its backbone on the inside. An aside: given that Pauling had discovered the alpha (single) helix structure for some proteins, why did he apparently jump to a three-helix structure for DNA? Why did he seemingly ignore the double-helix possibility? I have read all that I could find on Pauling's thoughts on DNA structure and am none the wiser as to why he apparently skipped a two-helix model.

Watson and Crick tried to develop a model consistent with the X-ray and chemical evidence. In particular, Watson, following Pauling, tried to build physical models of the structure, with metal pieces made to his specifications in the Cambridge metal shop. At first he worked with the phosphate-group backbone on the inside of the double helix. This configuration failed. Crick convinced him to put the backbone on the outside, as earlier suspected, with the base pairs residing on the inside. At first, Watson tried matching each base to an identical base on the other helix, ignoring Chargaff's rules. This structure, too, failed to meet the known requirements of the data. Watson then tried to link purines on one helix to pyrimidines on the other and vice-versa (see fig. 17.1). This plan ran into trouble with matching the known chemistry.

It was here that Jerry Donohue played his key role. He told Watson that the form for the structure of guanine found in chemistry books was likely wrong and that a different structure was correct. With this latter structure, Watson soon realized that the two sets of base pairs, guanine-cytosine and adenine-thiamine, had the same overall size and shape. He also realized that cytosine would be paired with guanine and adenine with thiamine in the structure of DNA, consistent with Chargaff's results. He was then able to get his model to match the data on the spacing of the nucleotides, on the pitch of the helix, on the connections of the structural elements, . . . in other words, success!

Of course, this result did not *prove* that his model represented the structure of DNA; it was just reasonably *consistent* with the then known facts. But it was a beautiful model; if nature didn't follow this model, it should have! The bases lined up pairwise, in accord with Chargaff's rules, with one purine matched to its pyrimidine mate via hydrogen atomic bonds (not to be here pursued in fundamental detail). These bonds could be relatively easily broken in a duplication scheme in which the two helices separated in the process of duplication. Such a process would take place in all cells, with each cell split into two during cell division, presumably also involving the DNA duplication, as discussed in the next chapter.

What about the competition? As far as I know, Pauling never again published anything about DNA structure. Wilkins and Franklin published only papers supporting the Watson and Crick model. In fact, Franklin herself came remarkably close to being the first to develop the correct model for DNA's structure. She decided, from her X-ray results, that the structure was helical, with the backbone on the outside and the bases on the inside. Then she got deflected from this track by an observational result on the A form of DNA that she found hard to understand. She puzzled over it, ignoring Crick's statement to her that this A-form result must just be an error of some kind, which she should ignore. However, she was too meticulous and thorough a researcher to do so. So, she only came close to uncovering the DNA structure, rather than uncovering it.

I emphasize once again that it is very hard to envision once you know the answer for the structure what it was like *not* to know the answer. It is easy, instead, to think, "Why, of course!" But in fact, it is no such thing. As proof, I merely note that the structure escaped many very brilliant people, and knowledge of the structure was achieved only by trial and error, plus much persistence, not simple insight as might seem logical once one knows the answer. I also add that the details of the story of this magnificent discovery are a *lot* more complicated and involved than could be indicated here, in terms of both the history and the science.

Watson and Crick completed their successful model in March 1953 and published two articles in *Nature* reporting their results, a very short one in April and a somewhat longer one in June. The first one had as its parting words a somewhat tongue-in-cheek statement: "It has not escaped our notice that the specific pairing [the double helix] that we have postulated immediately suggests a possible copying mechanism for the genetic material."

The second paper had the following prescient ending: "The phosphate-sugar backbone of our model is completely regular, but any sequence of the pairs of bases can fit into the [double-helical] structure. It follows that in a long molecule many different permutations are possible, and it therefore seems likely that the precise sequence of bases is the code which carries the genetical information."

How was the Watson and Crick model for the DNA structure received by the scientific community? The reaction was mixed. Some scientists, believe it or not, still held out for proteins as the prima donnas. Others noted: this suggested structure is just a model; it needs proof. Proof was indeed forthcoming, from more sophisticated X-ray crystallography. But it also involved modifications to the model, important to crystallographers but unimportant to contemporary biologists, and took more than twenty-five years, until 1979, to carry out successfully. By then, the critics had gone the way of the passenger pigeon due to the many other biological triumphs of the Watson and Crick model.

As for the lay community, Watson's mother, after hearing from him about his model, announced to her friends that her son had just found the secret of life. When Crick made a similar announcement to his wife, Odile, about his having made an important discovery, she later reported that she didn't believe a word, noting, "He was always making such pronouncements."

Tragically, Rosalind Franklin died of ovarian cancer at the age of thirty-seven in 1958. Crick, Watson, and Wilkins were awarded the Nobel Prize in Physiology and Medicine in 1962.

We move in the next chapter to a discussion of some of the elegant work in molecular biology inspired by this model of the structure of DNA.

NINETEEN

Properties and Functions of DNA

DNA, as Miescher first taught us, is in the nucleus of a cell (but, of course, only for cells which have a nucleus; only some do). When such a cell duplicates, how does the DNA duplicate, or replicate, the more commonly used word to describe this same phenomenon? As Gunter Stent, a biologist at UC Berkeley, summarized the proposals on DNA duplication by reputable biologists, there were three main possibilities, named semiconservative, fully conservative, and dispersive. Although the details were not spelled out, the ideas for the three were as follows: semiconservative means that the double helix separates and each of the two strands forms its own complementary mate, thus yielding two copies of the original double helix (under this scheme, further duplication of course follows the same pattern); conservative duplication means that a new copy is made of the original double helix, leaving the original one intact (in further duplications, the original double helix remains unchanged or, put another way, conserved); in dispersive duplication, the original DNA breaks into an unspecified number of various-length segments and each segment duplicates separately as in the conservative method, before reassembling themselves into DNA molecules of their original size. How could these three proposals be distinguished? In particular, what experiment(s) would allow us to make a proper choice among them?

Matthew Meselson and Franklin Stahl, then—mid- to late 1950s—both young researchers at Caltech, had an idea to apply a unity-of-science approach, although they of course also did not employ this term.

Meselson and Stahl used a heavy isotope of nitrogen, ^{15}N, a nitrogen atom that has in its nucleus an extra neutron compared to the contents of normal nitrogen, ^{14}N; see chapter 11 for a refresher on these properties of atoms and chapter 18 for the notation. By growing an organism, *Escherichia coli* (*E. coli*, for short), for several generations in a medium (source of food) grossly enriched in this heavy isotope, one can ensure that the organism's DNA has almost all of its nitrogen in the ^{15}N form. (*E. coli* is a so-called model organism, one that has desirable properties for experimental biology, such as short generation time, and is adopted for general use by different biologists for different research projects. One advantage is that all of the results can be pooled, and the community as a whole will then know more about this one organism and can see connections that might be important but would not be so easily discernible were the various results from the different studies to be for different organisms.)

Then Meselson and Stahl put these ^{15}N-enriched *E. coli* in an environment with only the ^{14}N in its food and discerned, after each successive generation for several generations, the relative amounts produced of the various DNA molecular weights. These results were then compared with the predictions from each of the three methods proposed for DNA replication. This comparison gave a resounding nod to the semiconservative means of DNA replication. For example, the ^{15}N-enriched *E. coli*, when allowed to reproduce in a food environment with only ^{14}N and no ^{15}N nitrogen, yielded no DNA with only ^{15}N nitrogen, as would be the case were fully conservative to have been the correct means of replication. Thus, semiconservative replication is not only what was expected, as noted in chapter 18, it is in fact how replication works.

The Code for Protein Manufacture

We are now convinced that DNA carries hereditary information and we now know—without knowledge of the details!—how it replicates. But what is its main message, and how does it convey this message? Its main message is likely how to produce proteins, which carry out all sorts of biological functions. A key question then is: How does DNA instruct the cell on the production of proteins? Likely the DNA makes use of a code somehow related to the base pairs. What do the words of the code signify? Likely amino acids, since these are the sole constituents of proteins.

The first proposed code was invented by George Gamow, of Big Bang fame, in the early summer of 1953, just following the publication of Watson and Crick's second paper. His scheme for the code was quite ingenious. However, it was quickly shown to not work either physically or chemically, so we won't pursue its details. Although Gamow's clever scheme was not viable, he may well have stimulated people's thinking about the issue of the code. One way this stimulation was instituted was via the creation in 1954 by Gamow and Watson of the RNA Tie Club, with its special tie for members. (RNA, and not DNA, was chosen for the title in part because RNA was felt to play a central role in the production of proteins from the original instructions for them in DNA.) This club had twenty members, then all elites of molecular biology—one member for each of the amino acids mainly used in biology—and four honorary members, one for each base. These members met, I believe, roughly twice per year for discussions whose goal was stimulation of each other's thinking. Once the problem of the code was solved, as we now describe, the tie club, all male members of course, shriveled and passed away.

In our thinking here, let's go back to basics: What is the minimum number of bases needed for a code distinguishing among these twenty amino acids? One base clearly doesn't work. Why? Because with only four different bases, we cannot code unambiguously for twenty amino acids; we could code only for four amino acids. What about a code of two bases? There are only sixteen possible permutations of two bases, as we can choose the first one in any one of four ways and the second one, independently, in any one of four ways, giving us $4 \times 4 = 16$ total possible permutations, and leaving us four shy of the needed minimum. Thus, the minimum number of bases needed for a code for twenty different items must be three: $4 \times 4 \times 4 = 64$ possible distinct three-base code words from four bases. Clearly, a three-base code is more than enough. We still have a basic (pun intended) question: If this is the way nature works, which sets of three bases code for which amino acids? Not an easy question to answer! It took about eight years and a heroic experiment to identify the code for just one amino acid. Actually, the very clever way it was done was the reverse, to find the amino acid that had a preselected code!

Side remarks are needed here before describing this experiment. Many scientists felt that the molecule RNA, a single-stranded relative of DNA and also a major cell constituent, is in several of its many different forms responsible for making proteins. RNA also has four, and only four, bases. But instead of having thiamine as one of the four, it has uracil, abbreviated "U." Why uracil replaced thiamine is not clear, just as the particular choice nature made of bases used in DNA and RNA is not clear.

On to the experiment: what Marshall Nirenberg, of the National Institutes of Health in the United States, and apparently mostly his postdoctoral student Heinrich Matthaei, neither a member of the tie club, cleverly did was to introduce a large amount of artificially created RNA into cytoplasm, the part of the cell outside the nucleus, extracted from a bunch of E. coli cells. This artificial RNA had only one base, repeated and repeated: UUUU . . . . (At this stage of technology development such a simple RNA was relatively easy to make but an arbitrary sequence was not.) By putting in a lot of this RNA, they overwhelmed any other RNA present in the cytoplasm, ensuring that any proteins produced would be due to the artificial RNA. They then loaded, separately, each of the twenty amino acids into its own container of cytoplasm, each of which contained the artificially created RNA. They found that only one type of protein was produced in substantial number; it was made of only one amino acid: phenylalanine. The conclusion? The code for phenylalanine was UUU, although how many successive Us were involved was not determined by this experiment; the number could have been three, four, or more.

Nirenberg presented this cleverly achieved result at the next opportunity, at an international conference, in Moscow, in the summer of 1961. Only about a half-dozen people attended the session at which he made his presentation. They were electrified. Word spread quickly among the conference attendees. Crick heard about it and had Nirenberg repeat his talk at the plenary session at the end of the conference. Everyone was impressed.

Thus was the first word decoded; the dam was cracked open. Clever experiments and theoretical work, primarily by Sydney Brenner and Crick, with some of this work having been done earlier than the Matthaei and Nirenberg experiment, led to other conclusions. A group of three bases (or, less likely, a multiple of three bases) codes for a single amino acid. The code is not of an "overlapping" type (for example, the last letter of the code for one amino acid does not also represent the first letter of the code for the next amino acid in the chain of the protein being described by the code). There are no special breaks, such as spaces between words or the analog of "commas," to show how to select correct triplets. The code is redundant: for some amino acids, more than one triplet of bases codes for the same amino acid.

Cracking the code for all of the amino acids took a lot of ingenious work by separate groups of very talented people and yielded the full (redundant) code, plus triplets signifying starting and (redundant) stopping of sets of triplets for the coding of an individual protein. This whole effort, stimulated by extreme competition among the various involved groups, was completed in 1966, only five years after the initial breakthrough. A Nobel Prize followed in 1968 for this work to Robert Holley, Har Gobind Khorana, and Nirenberg.

There are many redundancies in this code, as expected, given the sixty-four three-letter words that can be made from four bases, as letters. As prime examples, six triplets represent the amino acid leucine and a different set of six triplets represents serine. Also, three different triplets represent "stop," which signifies the end of the coding of a protein. How many triplets indicate "start"? Apparently only one: the triplet ATG in DNA denotes "start." It is the same triplet as for the amino acid methionine. How does the cell know how to interpret which use is which for ATG? I do not know the answer, other than to guess that, in the middle of the instructions for a protein, before a "stop" triplet has been encountered, the cell knows that "methionine" is meant and not "start." But if this triplet is encountered when not in the midst of instructions for a protein, then "start" is the interpretation made of this triplet. How one knows whether the base sequence is in the middle of instructions for a protein, or of something else, is yet another good question. Probably the correct answer: all proteins start with the amino acid methionine; it is, however, often excised, I assume, from the start of a protein by processes employed by the cell later in the production of the protein. Why, given the availability of so many three-letter codes, nature didn't use one — or several — separate triplet(s) of bases to signify only "start" no one yet knows, as far as I know!

The Central Dogma

Now we know the code that DNA uses to give instructions for the sequence of amino acids for the manufacture of each protein. But how do cells make the proteins from these instructions? Crick gave the name Central Dogma to the answer to this question. The critical component in this saga is RNA, which comes for this purpose in three main flavors.

The first is messenger RNA (mRNA), which copies the instructions for a protein from DNA by being built of the complementary bases (for example, C and G are complementary, as are A and T) to those between a "start" and the following "stop" triplet, except with U replacing T. Thus, for example, in mRNA, C and U would be placed when G and A had been in the corresponding places in DNA. This whole process is called transcription. The second is transfer RNA (tRNA), for which one such RNA exists for each of the twenty types of amino acids, binds to "its" amino acid, and carries it to the growing end of the protein at the ribosome, which builds the protein. This process is called translation. The third is ribosomal RNA (rRNA; ribosome), a complex RNA entity in which proteins are put together, seemingly in an assembly line fashion. Such a ribosome seems to be an extremely complicated structure; learning exactly, or even approximately, how the ribosome carries out the protein manufacturing process is not likely an easy task.

Does anyone really know all the details of this process? Probably not, though collectively the science community knows *much* more than I can present here. For example, on the DNA molecule the instructions for making proteins are composed of so-called exons and introns. There are many of each in the instructions for each protein, although more introns than exons. The introns are not copied to make the protein, the exons are, but different selections go into making different proteins. Thus, for example, a given set of base pairs on the DNA molecule can be involved in more than a single protein. How will new knowledge come about? I can point to one of many technical developments that may help: very high time resolution pictures of these processes, at the attosecond level (10^{-18} second level), which technology has now just about reached. How we will take such pictures and how we will digest the flood of results are good questions. Certainly part of the answers will be in improved and much faster computers and software to process the floods of data. The same type of comments will also apply to increased spatial resolution of the cell workings. More likely, major progress will flow from technical developments that I could now not even conceive of.

Beyond the manufacture of proteins, about which we still have many unaddressed questions, we have other fundamental questions to address. Here is a small sample: How is the orchestra of cell processes directed? What tells the cell when to replicate, and how is this information conveyed? How do materials in syntheses know where and when to go to the places they are needed and how to get there? What controls the rates at, and order in, which various cell components are made? How, in detail, does the supply side for a cell work? These questions may not now be on the tops of lists of priorities of molecular biologists but will likely be addressed in time and may well yield insightful answers—apple pie and motherhood . . . .

DNA Conundrums

We understand the function, but likely only partially, of only about 5 percent of the human DNA. This percentage involves mainly the instructions for making perhaps up to one hundred thousand different proteins in a human cell. (This number varies from cell to cell and, for the average cell, may well be uncertain by as much as a factor of ten or so—there is as yet no consensus among the experts, as far as I know.) These sequences of bases providing instructions for making proteins are called genes and represent what is usually considered the hereditary material of the DNA. (I say "usually" here because living organisms are sufficiently complicated that there seem to be exceptions to just about all rules.) Some parts of the DNA also control cell functions (for example, turning on and off various genes). What is the rest good for? There are many speculations, but I do not believe that we yet know. The situation is somewhat analogous to our not knowing much about what 95 percent of the mass-energy of the universe is made. Both are *big* puzzles; although the latter is much bigger volumetrically, the former likely has a far bigger direct effect on our lives!

There are also smaller DNA puzzles. Consider, for an example, the seemingly small issue of the total length of an organism's DNA molecule, albeit spread over many different chromosomes. A chromosome, by the way, is made up of part of the DNA of an organism, encased in protein. The whole DNA molecule is encompassed by the totality of an organism's chromosomes, with no repetition. Different organisms have different numbers of chromosomes; a human has twenty-three from each of their father and mother. Each of these halves is called the haploid form of DNA; together they form the diploid. One might naively expect this length to be correlated (positively) with our view of an organism's complexity. But nature has different ideas. Let us take four examples of organisms and their approximate numbers of base pairs in the (haploid) form of their DNA:

> onion, 20 billion
> lungfish, 133 billion
> salamander, 120 billion
> human, 3 billion

If these numbers are reasonably accurate, why, for example, we may well be tempted to ask, would an onion have nearly seven times the number of base pairs in its DNA as a human? When we thoroughly understand the answer, we will doubtless be much wiser, and perhaps in surprising ways.

We end this chapter with a calculation of the total length of DNA in our bodies, in a rough backwards equivalence to the pepper analogy we gave for the solar system in chapter 5. If you prefer to skip the details, please proceed directly to Step 3, below. On to the calculation, in the following three steps:

1. The length of the DNA in one cell equals the length of a DNA base pair multiplied by the number of DNA pairs, which is twice the haploid number (that is, the number of the mother's plus the father's contributions). Recall from the analysis of the X-ray images of DNA (see chapter 18) that each

base pair extends along the DNA axis by 0.34×10^{-9} meters; the number of base pairs is about 6×10^9. The product of these two numbers is about 2 meters.

2. The number of cells in your body is, roughly, the volume of your body divided by the average volume of a cell in your body. The former number is very roughly $0.1 \times 0.2 \times 2$ cubic meters. The latter number is, also *very* roughly, $10^{-5} \times 10^{-5} \times 10^{-5}$ cubic meters, with the quotient being about 4×10^{13} — that is, about forty trillion.

3. The total length of the DNA in a single human body is thus the product of the numbers from Steps 1 and 2, above, or about 10^{14} meters, which is nearly 700 astronomical units, comparable to the diameter of the entire solar system. Pushing further and taking the total length for all living human beings, about eight billion of us, we find a value for the total length of our collective human DNA of about seventy-five million light-years — over one thousand times the major diameter of our Milky Way galaxy!!

On that astronomical comparison, we leave DNA, to return to it in a different context in the next chapter.

TWENTY

The Origin of Life, the Tree of Life, and Samples of Evolution

We now turn to treat some other fundamental questions, the origin of life and the tree of life, as well as some important issues on evolutionary paths of two individual species. Of course, if truth be known, we do not yet know much in detail about the origin of life, and although we do know much about the tree of life, there is still much, much more to learn.

The decision here as to which specific topics to treat (and how deeply to delve) in such an incredibly broad field is certainly subjective, and mine, I admit, is also somewhat quixotic.

As we embark on these new topics we will keep in mind our previous prime lesson: a key to advances in science, and in other fields as well, is often the asking of the right questions.

The Origin of Life

Why have we waited until now to discuss the origin of life? Shouldn't this discussion have been first on our list of biology topics? Answer: We needed to set the stage with a proper background to allow us to know a bit about life before discussing its origin. We can cover only a few aspects of the search for the origin of life, but I hope enough to give a fair flavor of at least some of the attacks humans are making on this huge problem.

For our present discussion, we first consider how to define life. We do so by thinking about its characteristics; our goal is to concentrate on key attributes that allow a distinction between living and nonliving material. One small—definitely not unique—set that accomplishes this goal, albeit somewhat loosely phrased, is: store and transmit information (for example, DNA and the Central Dogma); "express" that information (for example, make "proper" proteins); and evolve (descent with "helpful" modification).

Next, we consider where life arose. One thought that has been around for well over a century is that life originated elsewhere in the universe and somehow arrived on earth. Of course, asserting that life came from afar only changes the question to another place of origin, about which we know much less. We will nonetheless return to this possibility in chapter 22; for now let us restrict our discussion to how life might have originated on the earth. Did it arise on land? In water? In the atmosphere? At an interface? The distinctions are not clear; for example, life could have been created at the interface of land and water with the atmosphere. But we should probably begin with the production of the chemicals essential for life as we know it, starting with basic chemicals, such as water.

An early experimental approach to answering this question of origin was undertaken in 1952 by Stanley Miller, then a graduate student in chemistry at the University of Chicago. He decided to test whether, given raw materials that we infer were in our atmosphere several billion years ago, chemicals such as amino acids could have arisen through natural processes. He started with the following constituents, in "appropriate" amounts: methane, ammonia, water, and hydrogen. (Oxygen was omitted because our best estimates have the atmosphere nearly free from oxygen 2.5 or more billion years ago; it was biological activity—plants exhale oxygen—that apparently led to the buildup of oxygen to its present level of about 20 percent of the earth's atmosphere.) Miller placed those materials in a sealed, otherwise evacuated glass container and subjected them to electrical discharges to simulate the input of energy into our young atmosphere from, say, lightning flashes. Energy is of course needed for any chemical build-up processes to take place. Miller ran the electrical discharges (electric sparks—artificial lightning) for a week. At the end, he identified the main chemicals then present in his apparatus. Lo and behold, he found the amino acid glycine and two forms of the amino acid alanine! Miller had been careful to avoid contamination, so that he could be sure that his original mixture of gases plus the input of energy from the electrical discharges had led to chemical reactions that yielded, among other molecules, these of the amino acids, albeit in relatively small amounts. Later experiments of the same sort yielded nucleotides, sugars, and more than twenty amino acids, including, I believe, all the ones used in biology. Alas, this new field of experimentally simulating the formation of chemicals that form the basis of life basically seems to have died with Miller's experiments, or at least hibernated. Why? I do not know, but I suspect that it may not have been clear how to do what was felt needed to be done next.

Even were all chemical ingredients in living organisms to have been produced in such simulations, would we have established life? Very far from it! A cell, for example, must self-assemble, and have a membrane that separates its inside from the outside, to protect it from harmful environments, and also to allow the cell to imbibe useful chemicals from its environment to grow and reproduce. The cell should also have a suitable source of internal energy, which we now know for us is provided mostly by the molecule adenosine triphosphate.

The formation and replication of RNAs and DNA must, somehow, be accomplished, as well as the manufacture of proteins. Then there's one of the conundrums: Which came first, proteins or DNA? Proteins do an enormous amount of biological work but cannot store or transmit information; DNA, by contrast, stores and transmits information but does no biological work—or so we believe. It's sort of like the classic question: the chicken first or the egg or . . . .

Perhaps the above conundrum is the wrong one to pursue. Maybe there's at least a third way: single-stranded RNA may have been first. One might need an entity, like RNA, that combines traits of DNA and proteins. It is involved in the critical machinery for replication and functioning of cells, as we saw in part in the previous chapter. An RNA molecule, with complicated folds and sites (locations on it), can catalyze chemical reactions. How can RNA be produced abiotically? An outstanding question—two senses!—that has spawned a whole field of study to address the means for evolving RNA in a test tube—in vitro (glass). The step from RNA to DNA would also need to have a pathway found. DNA has clear advantages over RNA: its double strand is stronger, its heat resistance is much greater (recall chapter 17 and Griffith's transforming principle), and it is more stable against chemical degradation (for example, from actions of acids and bases). We also, however, do not know that DNA evolved from RNA; although we think it likely, there may have been a different, albeit likely related, origin for DNA.

The building of multicelled organisms presents yet another host of problems to be overcome. The list of problems, in fact, goes on and on and on . . . almost beyond imagination.

Indeed, we have hardly even mentioned yet another possibility for the origin of life: underwater. As we learned in chapter 10, there is plenty of energy spewing from inside the earth at midocean ridges that could be used to create life. And we have relatively recently discovered all sorts of extremophiles that now live and thrive in those parts of the undersea environment. If life originated near such a deep vent, the processes for making the needed chemicals may—or may not—have been greatly different from those involved in Miller's approach.

Although an unsatisfactory ending, we stop this discussion about the origin of life with the statement that it is a very active field of research on many fronts, including, for example, the study of the possible origins of the membrane that separates the inside from the outside of a cell. There is still plenty to do—understatement of the decade!

The Tree of Life

What is the tree of life? It is a figment of our imaginations, an analogy, if you will, to the evolution of life from a presumed single origin to the bewildering biological diversity found today on earth. The roots represent the earliest forms, which branch out, with the leaves representing the current forms, in one realization of this tree analogy (or simile).

How do we construct a tree to represent the evolution of life? The answer: with difficulty, with aid from contemporary specimens and fossils, and with DNA analyses. We now think of life as having branched into three main domains: bacteria, archaea (many are extremophiles), and eukaryota (animals, plants, fungi, slime mold with cells each having a nucleus).

What distinguishes archaea from the other two? Like bacteria, archaea are single-celled organisms with no nucleus, but unlike either bacteria or eukaryota, archaea have a different composition of cell membrane that makes them more stable (perhaps relating to their including extremophiles); also, unlike the other two, archaea apparently produce methane.

Where do viruses fit among these branches? Biologists seem to ignore viruses in discussing the tree of life, presumably because they can't reproduce on their own. But they do just fine reproducing in what are considered bona fide life-forms. Since viruses reproduce only in other hosts, where would one place them on this tree of life? Not obvious. Perhaps at some point making them a fourth domain will be deemed a satisfactory solution.

Rapid increase in DNA sequencing of organisms is expected to vastly improve matters by reducing the number of somewhat arbitrary decisions made in tree constructions. This increase is expected to be due to both an increase in the speed of sequencing and a decrease in its cost. As an example of trends in cost, I note that for the human genome, the cost of sequencing it in the first twenty-one years of this century has dropped, I believe, by about a factor of nearly a hundred thousand! Of course, the past trend does not necessarily indicate the future trend. But such a continuation is not an unreasonable expectation (note the intentional double negative).

Evolutionary Paths: Polar Bears

Until very recently, little was known about the species history of polar bears. Why? The paucity of their fossil record: the environment in which polar bears live seems not to be conducive to the formation of fossils that are accessible. Polar bears spend virtually their entire lives on ice. Thus, after death they presumably sink in the ocean and are either eaten by suitable scavengers or perhaps form fossils that are not easily retrievable. There is also the problem of relatively small populations, which in the normal course of events would lead to relatively few fossils. Nevertheless, a few fossils have been discovered and analyzed. Most of the action in this field has, however, been with the analysis of the DNA of bears.

As indicated, we wish to understand the phylogeny—that is, the evolutionary history—of polar bears. Why? Outside of normal curiosity, we would like to know how rapidly polar bears adjusted to their exceedingly unusual environment. This knowledge could, for example, be a useful benchmark in understanding the pace of evolution in large mammals.

What have we been able to infer about polar bear evolution? First, all analyses agree that polar bears split from brown bears, evolutionarily speaking. These two types of bear now exhibit some profound differences in characteristics. Brown bears are omnivores, whereas polar bears are carnivores, subsisting almost entirely on seal blubber. Polar bears have amazingly high levels of body fat, 50 percent. By comparison, brown bears and humans have, roughly, about 20 percent body fat. The polar bear levels of cholesterol in their blood are astronomically high by human standards, 385 milligrams per milliliter (my value in the same units for comparison is about 180), yet polar bears do not seem to be afflicted by cardiovascular problems, which, by the by, poses challenges for those searching for underlying genetic causes of cardiovascular diseases in our own species. Also, the fur of polar bears is quite white, offering excellent camouflage in their snow-laden environment. What genetic changes gave rise to each of these different characteristics? For example, it is still not clear, I believe, what combination of genes is responsible for white fur. And how long it took for polar bears to acquire each of these characteristics as they split from the brown bear is an important unsolved issue; only an (approximate) upper bound is available for all of them. Can you think why?

The current global warming trend, apparently due at least in part to humans, is dangerous for polar bears: their habitat, ice in the Arctic, is disappearing rather rapidly in historical terms. They are thus subject to a relatively recent and dangerous predator—humans—and are harmed in other ways, too, such as by chemical pollution of various sorts, also due to humans. This combination is unprecedented and seriously threatening to polar bears, whose total population, although not well known, is estimated to be about twenty-five thousand.

Knowledge of their history might, just might, help in human planning of a strategy to prevent the extinction of polar bears, via, say, fatally low genetic

Figure 20.1. A mother polar bear and her two cubs posing for the photographer. Steven Amstrup/Polar Bears International.

diversity. A recent photo of contemporary polar bears shows one reason why we wish to preserve them (fig. 20.1).

To quantitatively trace polar bear history, the DNA trail is almost the only game in town, given the fossil paucity. The first such study of which I am aware was in the early to mid-1990s and used the mitochondrial DNA of polar bears. (Mitochondria are entities in the cell, outside the nucleus, that have their own DNA, generally much shorter than the DNA in the nucleus. These mitochondrial DNA are inherited through the mother; the number in a cell varies greatly, from zero in a human red blood cell to up to several thousand in a liver cell. The DNA of these mitochondria can be analyzed like the DNA in the nucleus to infer, at least in principle, the partial history of evolution of the host organism.) This examination indicated that the polar bear is most closely related to the brown bear, with which it interbred. This interbreeding was expected to occur, and most likely did occur, during interglacial periods, which provided opportunistic mating as the ranges of the two types of bear then overlapped. Some say that polar bears moved south during warm periods, seeking land, and others say that brown bears moved north as the climate there became more favorable to their lifestyle; maybe both scenarios occurred.

An important question: How old are polar bears? That is, when did they split from brown bears the first time, or, equivalently, when did they split from their last common ancestor? (The last common ancestor, or LCA, is, by definition, the most recent common ancestor; this terminology is used because, for example, the first common ancestor goes back to the original first life, which we assume was a unique event.) The answer to how long polar bears have existed can be approached from examining their fossils and/ or their DNA. With fossils, given the high reliability of radioactive dating coupled with the reasonable reliability of association of the fossil with contemporaneous surrounding material containing the radioactive elements for dating, the result has a distinct advantage over the inference from DNA analyses. And yet, the fossil, even if distinctive from that of the brown bear, cannot be reliably stated to be the first such. DNA analyses are also more subtle in the following way: one cannot just determine the differences between the DNA of a brown bear and that of a polar bear and apply a molecular clock reading to infer a splitting time for the two kinds of bears. Why not? We do not, as far as I know, yet have any evidence that there is a molecular clock that ticks at an even nearly constant rate. There is also the annoying complication that between an initial splitting and now, brown bears and polar bears have had unknown numbers and durations of interbreeding, which can result in flows of genetic information from polar to brown bear and/or vice versa. Moreover, for analyses based on mitochondrial DNA, there is the additional complication of the unknown effect of possible bias caused by sex-linked changes.

How can these problems be surmounted? Not easily, as I am prone to say. A whole host of sophisticated statistical methods have been developed, which treat these various aspects of the DNA sequences in different, seemingly plausible, manners. The overall problem is: different methods yield different results, some vastly different. Could it be that a potentially important effect has been overlooked? I refer, for example, to those of the possible change(s) in the three-dimensional structure of each section of DNA, due to changes in the base pairs at various locations in that section. Of course, as yet we have very little knowledge about, let alone understanding of, the effects of changes in the three-dimensional structure of DNA segments. Small changes in base pairs may cause relatively large changes in the three-dimensional conformation of the DNA. Note: this statement is purely qualitative! Guess why.

The analyses of the mitochondrial DNA yielded a range of estimates of the LCA with a maximum value of about 135,000 years ago. A later estimate, based on the analysis of the nuclear DNA, yielded a maximum of about 600,000 years ago (with stated high confidence intervals of about 200,000 years plus or minus). A recent estimate of the LCA jumped to between four and five million years ago. Is there a meaningful trend here? (Joke!) More importantly, is there a (good) explanation known for these differences in estimates and will we settle on a definitive answer? Stay tuned—watch the media; a good answer might be forthcoming in the next years (note the deserved qualitative aspect of this statement).

What then can we now conclude reliably about the time of divergence of the brown bear and the polar bear? In my opinion, not much. If we had a reasonably accurate value for this separation time, since we know with reasonable reliability the age at which the bears become sexually active, about four to five years, we could make a reasonable estimate of the number of generations required to produce the differences now seen between the two types of bear. We still await a reasonably accurate value for the separation time.

Interbreeding with brown bears, as in the past in interglacial periods, might once again be the savior of polar bears. Further, the prospects are reasonably bright for more insights into the past, independent of the present state of the analysis of polar bear DNA. Also, it is possible that more old fossils will be found and analyzed.

Evolutionary Paths: Whales

Whales are often huge, the largest living creatures. Some are even larger than were the largest dinosaurs. Whales are a different kind of creature, though, because they live in the sea. These dinosaurs of whales in terms of size can be visualized in part by comparing the size and weight of the largest whales with those of the largest elephants, some of those that inhabit

Africa: large blue whales can be up to about 30 meters long and weigh up to about 180 metric tons; the largest African elephants, from tip of extended tail to end of extended trunk, can reach about 10 meters in length and weigh up to a bit over 6 metric tons, about thirty times less massive than the most massive whales.

Despite their present ocean habitat, whales are mammals and not fish. The distinction is clear from many characteristics: whales, but not fish, give birth to live young and nurse their offspring; they have anatomical differences, too, such as lungs versus gills. Aristotle in the fourth century BCE is reputed to have pointed out, probably not the first to do so, that whales breathe air and nurse their young. Nonetheless, the classification of whales as mammals has not been universally accepted by the public. Exhibit A: Ishmael in Herman Melville's *Moby-Dick* (1851) refers to a whale as "a spouting fish with a horizontal tail."

Whence did whales arise? It is superficially, at least, somewhat ironic that land creatures are believed by some scientists to have evolved from creatures of the sea, yet whales seem to have been the end result of a reverse evolution—land to sea. Whales apparently evolved from protowhales that lived on land about thirty-five million years ago.

Filling in this interesting part of the tree of life was first tackled well over a century ago. Darwin even speculated about whales' origins in the first edition of *On the Origin of Species*. He postulated that bears, who went into the water to swallow hordes of insects, had been the land forebears of whales. As we will see, he was wrong on the animal that converted to maritime life but correct on the conversion. There still exist some fundamental questions yet unanswered about this evolutionary track. We describe this evolution, as did the original detectives, almost solely via the fossil record, not via DNA. This record is rather rich, despite the probable paucity of whale fossils as a whole, for reasons similar to those given previously for the near absence of polar bear fossils.

While most educated people know that whales, dolphins, and porpoises are mammals, almost no educated person knows how that happened. Specifically, what mammal(s) on land evolved into these sea mammals and via what evolutionary steps? The story starts about sixty million years ago.

How did paleontologists go about trying to understand this intriguing conversion? Here we will trace their approach for whales but not for the others. Why? The whale transition seems to have had the most effort spent on it, and in my opinion, this effort has led so far to the most fascinating results.

The paleontologist's most important source of information is clearly fossils. But to measure and study fossil bones, they must first be found. The world at the fossil-size level looks and is enormous: Where does one look? One needs to know where there are accessible rock formations that disclose layers of the "right" ages. And to find protowhales, one looks at land-sea interfaces from the period of interest. There is also a bit of luck involved in finding fossils.

A goal of paleontologists is to find complete fossil skeletons representing each of the major evolutionary steps. We have many samples of contemporary whale skeletons. Looking carefully at these bones, from head to toe, paleontologists try to identify those that show the least deviation from other, contemporary, land mammals. Retreating back through time, paleontologists noted many fossils, which they tried as best they could to put in correct chronological order, using the rules previously developed for such activities. In this way, the trail was blazed by paleontologists back to the fossil skeleton with hind limbs whose toes resembled those of ungulates (hooved, or hoofed, mammals, such as camels and hippopotami), with even numbers of toes.

Thus, the toes of these mammals can be seen to (sort of) resemble the corresponding bones of the whale. Other similarities are also consistent with this origin. We cannot explore this entire detective story in detail, alluring and illuminating of methods of evidence gathering and inference making as it actually is. Instead, we treat the evidence of mammalian origin of whales from ontogeny, and the basis for the association of whales with mammal forebears.

That whales originated as land mammals can be seen in their embryology. As the whale fetus develops (ontogeny), it goes through stages that repeat those of its evolution (phylogeny)—the origin of the famous phrase "ontogeny recapitulates phylogeny." In the case of the whale, for example, ontogeny shows the fetus having body hair, which disappears before birth, and having nostrils, which migrate from the usual place for mammals to the top of the head, the place of blowholes in whales. Thus, the ontogeny case for whales having developed from land mammals is rather convincing.

A breakthrough event in this saga occurred about two decades ago, in 2001. Philip Gingerich, a distant cousin of Owen's, and his colleagues discovered on the Indian subcontinent, in present-day Pakistan, an approximately forty-million-year-old, almost complete, fossil of a near-whale, a protowhale. This fossil filled a big gap in the evolutionary record, from which skilled paleontologists can draw rather remarkable and rather accurate conclusions, based on about two centuries of development of this field since Georges Cuvier invented it. Specifically, this fossil allowed the fairly definitive conclusion to be made that the whale's land-living ancestor was an artiodactyl (even-numbered [artio]; fingers or toes [dactyl]) ungulate (hoofed mammal), of which today's versions—that is, descendants—include camels and hippopotami, as also noted above.

What was the key evidence that clinched the case for even-numbered-toed ungulates, like the hippopotamus, being the land ancestors of whales? It was the comparison of ankle bones! The nearly complete skeleton found in 2001 by Gingerich and his colleagues showed the ankle bones to have a double-pulley aspect to them, a characteristic found only in contemporary even-numbered-toed ungulates (fig. 20.2). By contrast the ankle bones of a bear, for example, showed only a single pulley–like structure, not this double-pulley structure.

Uncovering this convincing—at least to paleontologists—connection was a major finding. The results were published by two separate groups a day apart in late September 2001, one article appearing in the prestigious United States journal *Science* and the other, a day earlier, in the equally prestigious, and worldwide perhaps even more prestigious, British journal *Nature*. A remarkable coincidence, right? Here is where the sociological, or human, aspect of research rears its sometimes, but rarely, less-than-principled head. The lead author of the *Science* article, Philip Gingerich, had early in June 2001 sent a draft of that paper for comment to a former student of his, then at a different institution. (We allow the curtain of charity to fall over the student's name for reasons soon to be clear.)

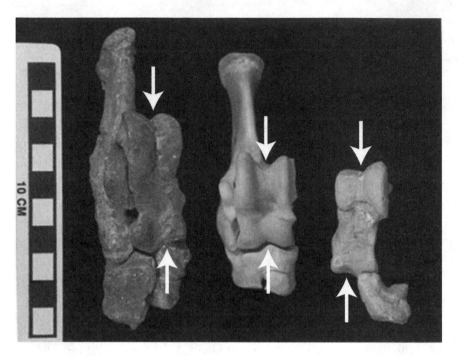

Figure 20.2. The double pulley–like structure (see arrows) is believed, based on the available evidence, to be unique to artiodactyls and the protowhale. Photograph by Philip Gingerich. Used by permission.

Gingerich, following the usual protocol, at the end of his group's *Science* article, thanked his former student for his comments on the draft. The former student, however, realizing the importance of this discovery to their common field of research, presumably then searched his collection of whale fossil bones, found some approximately matching those that formed the basis of his former adviser's paper, and immediately wrote a paper drawing most of the same conclusions, and submitted it to *Nature*, probably impressing upon the editor, either directly or indirectly, the desirability of rapid publication in view of the importance of the result. *Nature* complied, and as a result, the *Nature* version of this key paleontological result appeared one day earlier than the *Science* version. (Both *Nature* and *Science* are weekly publications, with the former appearing on Thursdays and the latter on Fridays, hence the one-day difference in publication date.) The one-day difference may seem to be trivial, but not necessarily so the implications, which can be more profound: the earlier publication usually gets most of the publicity and, in the eyes of the scholarly community, often most of the credit.

Is the tale told above reliable, or could there be some deep misunderstandings that vitiate the main storyline? Of course, it is hard to be sure of happenings that are largely inferential. What objective evidence is there in this case? A lot: the *Science* article was submitted on June 28, 2011; the *Nature* article was submitted on August 10 and accepted for publication on August 28, 2011. An eighteen-day turnaround from submission to acceptance is a conspicuous anomaly, as is the submission-to-publication interval, suggesting editorial intervention. The *Science* article contains an acknowledgment by the senior author to his former student, the lead author on the *Nature* paper, for his having commented on the draft of the *Science* paper. The *Nature* paper, by contrast, included no reference to the thesis adviser's related work but did include a "note-added-in-proof," meaning it was added after the paper was accepted and while its page proofs were being read. The note stated, "Close cetacean-artiodactyl ["Cetacean," by the way, means a mammal, such as a whale, that lives in the ocean] relations are also implied by protocetid fossils in an upcoming paper," with the reference being to the paper by Gingerich and colleagues to appear in *Science*. All well and good, except that the note-added-in-proof was implied to have been written well after the paper was submitted, which submission was well *after* the author had read the draft of the *Science* paper and well *after* that paper's senior author had received the comments back from his former student, the *Nature* paper's lead author. It is thus hard to conclude other than that the note-added-in-proof was a disingenuous statement. Perhaps not too surprising, the former adviser and his former student are no longer on good terms. (By now, too, most people in this general field know what happened.)

To the best of my knowledge of the sociological aspects of this priority issue, the story ends here. It is a story largely of one aspect of human behavior. However, to put it in perspective, I mention that I have published more than four hundred scientific papers, with a total of some 250 colleagues over the years, and from this type of shenanigans, I was victimized only once, albeit in an even more egregious way.

It is worth adding a science-related point. The specimens described by Gingerich and colleagues were nearly complete skeletons, found in Pakistan, showing direct association of the ankle bones with those of ungulates. His former student's material was entirely isolated, individual bones collected earlier from a bone bed that mixed in skeletons of animals of various different types (Gingerich had had such specimens, too). The linking of the whale ankle bones to those of (modern) artiodactyl land-dwelling mammals, however, was not done earlier by either Gingerich or his former student (or anyone else, for that matter) because it could not be done easily without the skeletons Gingerich and colleagues found in 2001. (I think here of the analogy with Greek and Egyptian hieroglyph texts lying on separate stones that couldn't be linked until the Rosetta Stone itself was found—and then deciphered only by inspired work.)

The most recent spectacular discovery in this area of which I am aware was again made by Philip Gingerich, in 2009, and was of two nearly complete skeletons of whale predecessors that had nearly evolved to the status of whales and were also found in Pakistan. This discovery was of two side-by-side fossils, with one having another creature inside it! We can rule out a creature that had been eaten, because it shows no signs of teeth marks as it would were it to have been ingested. The only reasonable interpretation is of a fetus, just before birth. Can you even imagine the excitement engendered by this discovery?

In fact, the fetus is aligned to be born headfirst. Its bones closely resemble those of whales. Moreover (weak argument?), the other fossil creature is apparently bigger, indicating that it was likely a male, in keeping with the relative sizes of contemporary whales of different sexes. These whales have structural signatures of being nearly fully aquatic. But there was a notable difference: fully aquatic whales give birth in the ocean, and the tail emerges first. For births on land, as we know from mammals, the head emerges first, as it seems for this fossil with a near-term fetus (unless a last-minute turnaround was part of the fetus's birth process).

Why might whales living wholly in the ocean give birth to young with tail, and not head, emerging first? We do not know reliably. One plausible theory: on land, mammals have the best chance for survival if they can breathe upon first emerging from the womb; if the feet were to emerge first and to get stuck, even briefly, the newborn might suffocate, since the amniotic fluid has already leaked out and the fetus is on its own. But it is still attached to its umbilical cord so where's the problem? In the water, were the head to emerge first, the newborn whale could drown before it was fully born. But why wouldn't it drown right after birth? At least for some whale species, the newborn calf rises to the surface within about ten seconds to breathe. Relatedly: How long can whales go between breaths of air? Some mature whales can remain submerged for almost two hours on one breath of air. How do whales breathe when sleeping? Apparently they remain half awake and do rise to the surface to breathe as needed.

We address one more broad and basic question: Why did artiodactyls take to water in the first place, and why did almost all other land species remain wholly on land? The answer is that we don't yet know. But we can speculate. One possibility is that artiodactyls took to water to avoid land predators. Crocodiles, then apparently well developed, could have been an impediment to the artiodactyls taking to water (obviously, though, not an insuperable impediment). Another possibility is the need for (more) food; this is unlikely, since evidence seems to indicate that food was then plentiful on land. Or they may have needed to avoid competition for food. Maybe, but evidence is not clear. And why not other species? Perhaps locations of earliest fossils provide clues; not obvious.

On that uncertain note, we move to our last, more hopeful, sentence of this section: we can expect future fossil discoveries, and improved DNA analyses, to better fill in this emerging evolutionary saga and, who knows, perhaps even uncover dramatically new and very intriguing aspects.

TWENTY-ONE

Tardigrades, the Applications of DNA, and the Human Spread across the Globe

This chapter could also have been titled "Miscellany." Here are a few subjects that don't really fit together but that I wished to include. Of course, of the applications, we can only treat a small sample, but I have tried to give a somewhat representative glimpse of their breadth.

Tardigrades

Tardigrades ("slow steppers") were discovered in 1773 by German zoologist Johann Goeze and named three years later by Italian biologist Lazzaro Spallanzini. In the classification of life, tardigrades have a whole phylum devoted to themselves. What, you may wonder, makes them worthy of discussion? It is their really incredible survival characteristics, some of which we now list. They've survived in outer space (outside a spacecraft) for several days at a time and at temperature extremes ranging from 1 Kelvin to 420 Kelvin for up to several minutes at each extreme. They've been frozen for more than thirty years and then been revived and been able to reproduce. They have been dehydrated to less than 3 percent water and survived well (a human is normally around 60 percent water). They've survived pressures of up to 6,000 atmospheres and ionizing radiation of 1,000 times a lethal dose for humans. Tardigrades, despite these extreme characteristics, are not considered to be *extremophiles*. Why not? Claimed answer: they do not *exploit* extreme conditions; they only *endure* them. Now that you have

Figure 21.1. A tardigrade, about 0.5 millimeters long. Copyright Eye of Science / Science Source.

been introduced to them, are you not amazed that most people, perhaps before now including you, don't know that they exist?

Tardigrades have been around for more than 530 million years; there are now over a thousand species of them. Tardigrades live all over the world—from the tropics to Antarctica. They travel over long distances using the wind, and over short distances, they may, according to a very new Polish study, make use of clinging to snails. They are oviparous, with their eggs hatching in as little as four to five days, and mostly about fourteen days, depending on species; they eat algae, small worms, and other such. Their lengths range from 0.1 to 1.5 millimeters, and their lifetimes, when not in a cryptobiotic state, range from three to four months to two years, depending again on species. In the picture of a tardigrade shown here, note the apparent absence of eyes, ears, and nose, which may lead to a lack of certain specialists among tardigrade physicians (fig. 21.1). Their DNA base pairs number 75 to 800 million, depending again on species, and they possess a protein (Dsup, or damage suppressor), which protects them against damage from X-rays; they also possess more copies of a DNA repair gene than any other known animal.

Tardigrades are expected to be able to survive all manner of extinction events and have done just that so far. (Of course, all creatures alive today have ancestors who, collectively, survived all extinction events at least until they reproduced.) It would be well were we to understand tardigrade tricks better so that someday we may be able to transfer some of them, and/or others, to humans! In fact, this pursuit has become an active area of research in some biological laboratories.

Overall, we have to ask the question: How did Tardigrades evolve to possess such incredible survivor characteristics? In particular, given that most of their special properties seem to be far more than is needed to survive any conditions encountered on earth, how did natural selection lead to tardigrades evolving these characteristics?? Good question. I do not believe there is, as yet, a demonstrated good answer—at least I am not aware of any.

An Evolution Experiment

Why have there seem to have been only very few experiments done to test evolution? The time required for evolution by natural selection to change species is usually measured in millions of years; a human lifetime is no more than about one hundred years, thus likely over ten thousand times too short to allow a single person to carry out a meaningful experiment in evolution, one in which a species change has occurred. There is one current exception of which I am aware to this lack of long-term experimentation on such a fundamental question. As is to be expected, it involves an organism with a short generation time, the so-called model organism the bacterium *Escherichia coli*, or, as it is universally known, *E. coli* (both noted above). It is a single-celled bacterium, about 1 micron long, and generally harmless to humans. Indeed, each of us has huge numbers of *E. coli* cells in our intestines. (Sometimes, though, *E. coli* is involved in food poisoning and other untoward events in humans.) Its generation time is well under an hour, about twenty minutes. Richard Lenski, now at Michigan State University, started this ongoing experiment in 1988. His goal was to see what evolutionary changes might occur were he to maintain the experiment for a very long time.

The experiment is rather simple, but very clever, in conception. Lenski took one *E. coli* cell and let it grow into many identical daughter cells. Using those daughter cells, he started twelve approximately identical populations, each in its own small flask containing 10 milliliters of a broth with only glucose to eat. Each flask was separately swirled to ensure that it had enough oxygen, and each was kept at the constant temperature of 98.6° F. In each population, the cells competed for the glucose, which ran out in about six hours. The cells then had to wait eighteen hours for the next feeding. Each day, the twelve populations, each in its own, separate flask, would have 1 percent of its *E. coli* extracted and placed in 9.9 milliliters of uncontaminated, fresh glucose-infused liquid, so that each of the twelve populations started at the same size and had a fresh six-hour food supply. After every five hundred generations, samples from each of the twelve populations were frozen and kept, separately, in a super freezer at a temperature of −112° F. Samples from these frozen sets of *E. coli* could be thawed, as needed, to go back in time and redo parts of the experiment, with the thawing process not disturbing noticeably any of the *E. coli*. Simple in principle, and rigorous in practice, this experiment was conducted to ensure that these same conditions were met, generation after generation, day in and day out, weekends and holidays not excepted.

What was found? After about ten thousand generations, the experimenters saw that all twelve populations were growing faster than at the start and that the cells were each generally larger. Moreover, all twelve populations lost their ability to grow in solutions with a sugar, D-ribose, an ability that their ancestors had had at the start of the experiment. Thus, the twelve showed parallel evolution—that is, evolution in which organisms that are not closely related independently evolve similar traits as they adapt to the same, or very similar, environments. Given the changed conditions, the evolution of the *E. coli* proceeded in a direction to adapt to the new conditions. It might seem strange that losing an ability could be adaptive, but if costly, losing an ability could be helpful. For example, maybe the *E. coli* were able to save energy by not building proteins that allowed D-ribose to be digested.

The most fascinating finding to date is that one of these twelve colonies, after about thirty thousand generations, was discovered to have a new ability:

these *E. coli* could thrive by eating a product that was inedible by other *E. coli*. And this new property was inherited. How did it come about? The answer is not yet totally clear, but presumably more than one (rare) mutation made these *E. coli* ready to adapt to eating this product, and adapt they did. What is the food that the mutations allowed *E. coli* to eat and under which environmental conditions? Two questions in one, which we'll answer sequentially: ordinary citrate, which was part of the liquid containing the main source of food intended for *E. coli*. Citrate—salt of citric acid—is usually not accepted into *E. coli* when the environment contains oxygen, as in this experiment. But the mutations were such as to—very improbably!—allow the *E. coli* to ingest citrate, as it would were the environment anaerobic— that is, without oxygen. Thus was made this major discovery in evolution.

How reproducible was this bit of evolution? Clearly, since it arose in only one of the twelve original colonies of presumably identical *E. coli*, the answer is "not very." One of the issues about evolution is to what extent evolution would be reproducible, given the same environment, and to what extent would it be unpredictable (random). This result of the evolution experiment weighs on the unpredictable or random side. On the reproducible side was the earlier result (see above) that for all twelve of the original populations, as time went on, the individual *E. coli* grew faster and larger.

How much longer would the experiment need to be continued before a clear change in species would be observed? No one knows, nor would there necessarily be agreement that a new species had evolved. But such a result, were it to happen, would be a crowning achievement. Why? If for no other reason than it would put a serious dent in the argument of evolution deniers that with all the undeniable changes produced by human breeding of animals and plants, never has a new species resulted. Horses only beget horses, dogs, dogs, and so on.

How long will this *E. coli* experiment continue? By 2021, it had passed the seventy-thousand mark in generations, but no one knows how much higher the count will go. Although Lenski will retire someday, he has trained many very competent successors. Further, other laboratories have now taken on these experimental evolution projects; it's a vibrant and expanding field of research. How long, though, will funding continue for such long-term projects is another question . . . . We can hope. These experiments are certainly addressing fundamental issues in evolution.

DNA Fingerprinting

We all take DNA fingerprinting for granted; it's old hat. But someone had to invent it. The inventor, (now Sir) Alec Jeffreys, developed the idea and the technique in 1983–1984. You might think, what's the big deal? You just sequence the base pairs in an individual's DNA and the game is over. Everyone's DNA is unique, just as everyone is different from everyone else, including, it seems, identical twins (and triplets, and . . .). Back then, and even now, the cost of such sequencing, and the time required, made and make it prohibitive to determine the entire sequence of DNA in a routine manner. Cleverness was called for.

In Jeffreys's time, it was well known that most of the total of six billion base pairs in the human DNA spread over each person's forty-six chromosomes were the same across our species; only a relatively small percentage, about 0.1 percent, varied from one individual to another. Of those that vary, two basic types are quite useful for our present purpose: short tandem repeats (STRs) and single nucleotide polymorphisms (SNPs). STRs are repetitions, one right after the other, in the DNA of short segments of base pairs. For example, CTGCTGCTGCTG is a short segment of three bases—C, T, and G—on one half of a DNA double helix, and it is repeated four times. (On the other half of the double helix are the corresponding bases, GAC.) These STRs vary from individual to individual in the number of times this segment is repeated and are apparently always in the non-protein-coding parts of our DNA. In contrast, SNPs are *single* sites in the DNA at which the base occupying that site varies from one person to another. For example, you might have a C at that site and I a T.

Why are these aspects of DNA useful? Since they vary from individual to individual, they can be used to distinguish among individuals. Well, you might say, "With an SNP, there are only four different possibilities for the nucleotide occupying that site, either A, T, C, or G. That's not very useful." True. But if you use a large set of such SNPs (and human DNA apparently has several millions of them all told), then you have a very useful discriminant indeed. Suppose that for each such site the probability is equal that an individual selected at random will have any particular one of the four nucleotides at that site. Suppose further that you select a modest number of such sites, say fifteen. Then the probability that any individual selected at random will have a specific set of nucleotides at those fifteen SNP sites is only $(1/4)^{15}$—a very small number, about 10^{-9}. Thus, if we use fifteen SNPs, we can be assured that, on average, only one person in a billion will have those same nucleotides at those locations. Of course, we would have to examine DNA samples from a reasonably large number of people from the population that we are interested in to check on the randomness assumptions. If the results were found not to be consistent with these assumptions, we would have to correct our probability calculations accordingly to know how well we will be able to distinguish one individual from another in that population. Or, for example, we could add more SNPs to our requirement.

Jeffreys proposed using these techniques to identify people with a high degree of certainty by sequencing only a select, tiny portion of their DNA. For example, depending on the circumstances, there might be only a small fragment of DNA available of the person to be identified. To carry out the analysis, say, of a set of SNPs in this fragment, it is helpful, nay essential, to be able to produce—and work with—a large number of exact copies of this fragment. Why? Most of the present methods for reading a DNA sequence rely on substituting fluorescent chemicals for the bases, a different chemical for each different base chemical. With only one or a small number of DNA samples, such reading is currently very difficult for humans or machines to read reliably.

A method to accomplish the needed duplication task was invented at about the same time as Jeffreys came up with his application. This technique is called polymerase chain reaction (PCR) and was invented in 1983 by Kary Mullis. It allows one to accurately reproduce DNA fragments many times rather quickly. Now one can reliably reproduce of the order of one million copies of lengths of DNA, from about forty base pairs to two thousand base pairs long, in minutes! For longer segments, special—slower—techniques can be used, with the present length limit being about fifty thousand bases.

How does this incredible technique work? The underlying idea is that DNA separates easily into single-strand components at high temperatures, about 95° C, just a bit lower than the boiling temperature of water, 100° C. After it reaches this temperature, and the DNA double helix separates into two strands, the DNA is cooled to a temperature of about 60° C, which though still very hot by our personal standards is low enough to allow two special primers to attach near the ends of the DNA strands to be duplicated. These primers are each of order twenty base pairs long and are made — somehow! — to be complementary to the base pairs at the ends of the strand to be duplicated. The primers are needed for the critical step, the one in which DNA polymerase (an enzyme; see chapter 17) causes the DNA single strands to duplicate, given that the environment has all the necessary ingredients, such as the needed nucleotides. This enzyme must survive at high temperatures and survive temperature cycling. Mullis visited the hot springs at Yellowstone National Park and found a bacterium — *Thermus aquaticus* — that possessed such a temperature-tolerant enzyme; it could also withstand the temperature cycling required for the PCR mechanism to work.

Modern versions of this process use, say, two ovens, one set at the high temperature, one at the low temperature, both with its needed ingredients and located next to each other. The material to be duplicated is then cycled between the two ovens for the seconds needed in each to carry out the required function. Each cycle doubles the number of DNA fragments. It takes twenty doublings to increase the duplicated number of DNA fragments to a million ($\sim 2^{20}$) copies. That's why it's called a chain reaction. If all of this sounds complicated, that's because in its details it is — but it does work and works exceedingly well. In fact, this PCR works so well that it has been used regularly for some time now by most experimental molecular biologists.

Back to Jeffreys: one of the first applications of his technique was to forensics, mainly to identify people thought to have been connected with a crime. For this application, the courts had to accept this newfangled evidence. All things considered, this type of DNA identification was accepted, legally, throughout Western countries, and some Eastern ones, after only a relatively short time of a few years. Now, over three decades later, we find that there is a thriving group, the Innocence Project, whose sole purpose is to investigate criminal convictions of people, by using available DNA evidence to check on the proper or improper identification of the convicted individuals. After about twenty years, the project has succeeded in getting the convictions of more than three hundred people overturned, while others have been convicted who were not touched in the original investigations and prosecutions for the crimes in question.

Studying the Black Death via DNA Analyses

At least three large human pandemics of the plague (the Black Death) swept through parts of the earth in the past nearly two thousand years. A Justinian plague hit Europe between the sixth and eighth centuries; the "big one," also mainly in Europe, took place in the mid-fourteenth century; and a late nineteenth-century pandemic mainly affected China and India. All were different, and in different ways. Our central question is: How do DNA analyses enlighten us on these events?

There are three recognized forms of the plague, and each has its symptoms (there is some overlap). Symptoms of the bubonic plague were painful and enlarged lymph nodes (called buboes, hence "bubonic"), chills, headache, fever, and general weakness. Sufferers of the septicemic plague experienced fever, weakness, abdominal pains, chills, and tissue bleeding (which may cause dying tissues to appear black and hence may be the origin of the term "Black Death"—nothing is for sure, however, in this game of look-back). The pneumonic plague caused chest pains, shortness of breath, cough, fever, chills, nausea, vomiting, and diarrhea.

The common belief is that rodents (especially rats, which stow away on ships and thus travel far) carry the disease and are bitten by fleas, which in turn bite humans, who then develop the disease. Many people, albeit an apparent minority, disagree, believing that fleas could never have spread the disease as rapidly as it apparently spread and therefore that the disease must have spread from human to human through the air. As far as I know, this latter form has not been demonstrated or disproven, but it is now widely believed that the pneumonic form, at least, can be transmitted between humans via aerosols of infective droplets.

The basic cause of the plague, most relevant scientists now agree, is the bacterium known by the friendly name *Yersinia pestis*, after Alexandre Yersin, the physician who identified this bacterium in 1894 from victims of a plague outbreak in Hong Kong. The *pestis* part of the name is self-explanatory. Whether, and if so how, this bacterium may have evolved from the Justinian period to modern times is still a matter of mild debate. In the Middle Ages, of course, bacteria were unknown. So other causes had to be conceived. I mention two: the king of France apparently was told by his subjects that the cause was a "great pestilence in the air," due to a conjunction in 1345 of three planets, proving that at that time astrology was alive and well, if nonetheless as useless as ever. The other was less innocuous: Jews did it by poisoning wells. Pope Clement VI pointed out in a bull that this attribution was preposterous, because Jews, too, were dying of the plague just like Catholics. No matter, facts, then as now, don't often alter people's minds. Jewish communities were attacked and hundreds apparently destroyed. Anti-Semitism, like astrology, was alive and well.

For all three types of the plague, the disease seems to take from two to seven days after infection to show symptoms and a roughly equivalent period until recovery or death.

Some historians estimate that the mid-fourteenth-century resurgence of the Black Death killed about one-third of all Europeans in about three years! Perhaps fifty million people in total perished; this number is, however, *very* uncertain: no good statistics were gathered, especially in view of the horrendous numbers of people dying; burying them may have been about all that survivors could accomplish. In fact a new study concludes that the death rate varied significantly over various parts of Europe and that a much smaller total number of people were killed. No matter, this pandemic apparently does make our two twentieth-century world wars and our twenty-first-century Covid pandemic look somewhat anemic by comparison, especially in terms of the fraction of the total population that evidently perished.

In the 1980s, remains of four bodies of plague victims in 1348 were dug up from a cemetery near the Tower of London. Several decades later, Kirsten Bos, Johannes Krause, and their colleagues teased out genetic material from the inner pulp chamber of teeth from the remains of these four individuals, with the goal of separating out the Y. *pestis* genome from the plague bacterium. They had many difficult technical problems to resolve, two serious ones we stress here: first, the DNA was found only in myriad separate, small strands; and second, these strands were overwhelmed by the DNA of the blizzard of bacteria that had infiltrated each body after death. Yet these authors managed to piece together a complete DNA for Y. *pestis*, the first such reconstruction of an entire genome (the complete DNA of an organism, and a word that I use with abandon hereafter!) of a disease from skeletal remains. Later, similar reconstructions were done from dug-up remains from contemporary cemeteries in other parts of Europe. These results virtually proved that the Black Death of the Middle Ages was caused by the Y. *pestis* bacterium.

How much, if at all, did the Y. *pestis* genome that caused the Black Death change from the mid-fourteenth century to today? To answer this question requires a comparison of the older genome with the modern one(s) and, as a check, the comparison of several older ones with one another and of several modern ones with one another. What comparisons were done? All I know is that the Y. *pestis* genome has one chromosome, with its DNA extending 4.6 million base pairs in length. The comparison between the modern and the fourteenth-century genomes shows only ninety-seven differences, with only twelve of these occurring in genes—that is, in the base pairs coding for protein manufacture in the bacterium. These numbers are very small. What are the differences and to which aspects of the bacterium's effects do these differences in base pairs relate? How varied, for example, are the genomes of different contemporary Y. *pestis* bacteria one from the other or of different fourteenth-century Y. *pestis* bacteria from one another? I do not know but feel quite sure that some scientists have addressed, or soon will address, these issues, or have a good reason not to!

Progress has also proceeded on another front: exhumed for DNA analysis were teeth from two individuals who succumbed to the Justinian plague, which apparently killed "many" tens of millions of people between 500 and 800 CE. (Justinian was the emperor of the Byzantine Empire when the plague there, in 541 CE, was apparently at its peak.) These two individuals had been buried in Bavaria, Germany. The analysis of the DNA in their teeth confirmed that Y. *pestis*, or a very close relative to the version involved in the Middle Ages plague, was the culprit for the Justinian plague about eight hundred years earlier.

What makes one version of Y. *pestis* virulent for humans and others perhaps not virulent? To my knowledge, there is little to no knowledge on this important point. Perhaps relatedly, why does a pandemic suddenly break out at one time and then not again for about eight hundred years? Again, no one knows. So, are we apt to have another Y. *pestis* pandemic? Scientists say that such an outbreak is unlikely now, basically for two reasons: first, hygiene is now much better and we don't have (infected) rats running all over the place (some might dispute this assertion!); and, second, we now have antibiotics, which provide efficient cures for Y. *pestis*. Thus, you may rest easy, but perhaps not *too* easy.

Let us end this section with a remark on an intriguing relatively new area of study: the role of the Justinian plague in the fall of the eastern Roman Empire!

CRISPR-Cas9 and Its Impact

These days CRISPR-Cas9 is on everyone's tongue. Well, maybe just the CRISPR part and perhaps not on everyone's tongue. But certainly it's close to, or on, the tongues of many biologists. Why, you may wonder, if you don't already know. It's the latest, hottest tool to be developed by biologists and has great promise to have profound applications for humans. Many of these ways have likely not yet even been conceived.

So, what exactly is CRISPR? The answer to that reasonable question is not simple. Let us approach it in two stages, giving first a rough idea and then an historical development, up to the minute, so to say. But beware: this discussion is not possible to fully present and, hence, not to be fully understandable; you should probably be satisfied to get the overall idea, recognizing that I have not explained—nor do I know them!—enough of the details to allow full understanding.

First of all, CRISPR is a molecular system that evolved in bacteria and has been coopted by humans as a tool for rewriting DNA in any living organism. It is also an acronym standing for clustered regularly interspaced short palindromic repeats—just so you know, not that knowing, per se, will help your understanding. Cas9 is a particular protein that works in concert with CRISPR to carry out the main function: in nature this function is to stop infection by cutting the DNA of an invading virus. In the gene-editing application described below, this function is editing DNA by cutting it at a specific site and triggering—in some way!—the host cell to change the sequence at precisely that location, replacing it with a different "hand-picked" part. From this exceedingly brief description, one can readily imagine some of the potential applications and corresponding ethical and moral dilemmas thereby posed.

Now for the very brief history. In 1987, the CRISPR phenomenon was discovered in a bacterium by a Japanese scientist, Yoshizumi Ishino. Its function was unknown. All that he and his colleagues knew was that there was this strange sequence of bases in the DNA of the bacterium *E. coli* that repeated eleven times with spacers of bases in between repeats, with each spacer being of length between thirty and thirty-three bases. The CRISPR array itself was made of DNA pieces each of length of about thirty bases (a near equal number: coincidence or . . . ?).

This CRISPR played a key role in the immunity of this bacterium from the deleterious effects of invasion by viruses. In fact, it was discovered that the spacers between the repeated sequence of CRISPR were extracts from the DNA of the virus invading the bacterium. The CRISPR with the DNA from the virus conferred this immunity on the *E. coli* bacterium: the spacers that complement the parts of the virus DNA that they represent serve as guides for the Cas9 protein to cut the viral DNA at the correct place and thus to provide the immunity, via details unclear to me.

Then within about a decade, scientists did experiments that, cumulatively, demonstrated another key property of CRISPR sequences. In combination with Cas9, and guided by a certain type of RNA molecule (molecule produced from the virus DNA extracts in the CRISPR), CRISPR had the amazing property that it could be programmed to cut, or break, DNA at a preselected spot.

So, for organisms whose cells have a nucleus (for example, all mammals), this CRISPR-Cas9 combination with its guide RNA had to be introduced into such a cell. The target cells would, for example, be in a dish with suitable nutrients. How to accomplish this introduction to allow this combination to efficiently carry out its cutting job was a problem solved rather quickly. Many copies of the combination were made first. Once inside a cell, the guide RNA would bind to Cas9 and seek out its base-complementary sequence on the DNA in that cell, bind to that sequence, and—in some way—allow Cas9 to make its cut.

Efficient construction of the desired combination of the RNA molecules was developed in 2012 by Jennifer Doudna, Emmanuelle Charpentier, and their colleagues, allowing simple use of the procedure described briefly above. More important, it was their original idea to use the CRISPR system for gene editing. The changes they introduced also seem to have made it much easier, faster, and cheaper to use this technique. Thus, two other techniques that accomplished the same result were left by the wayside, while this one captured the enthusiastic attention of the relevant community of biologists and became the standard tool in the relevant laboratories that deal with these and related issues by about 2015. For their work on CRISPR-Cas9, Doudna and Charpentier received the 2020 Nobel Prize in Chemistry.

How reliable is this technique? That is, how often does it not work at all or cut the wrong part of the DNA? My understanding is: very reliable with a good guide, which is usually if not always obtained and yields error rates under 0.1 percent and undetectable with current technology. There is—at least!—one more relevant question: How does the cell repair its DNA after excision of the targeted stretch of DNA? There are several ways that the cell can do this reattachment, but it is not yet reliably predictable, or controllable, which method the cell will use in any given situation. To my knowledge, finding this answer lies in the future.

In the long run what is this technical capability expected to be good for? It certainly sounds neat, but how will we use it to increase our basic knowledge of biology and to improve, say, the human condition? All biologists, and many others, are very eager to realize the prospects, although it will likely be many years before most are realized: there are ethical and legal considerations galore that must be dealt with, many of which are best done on an international level. Aside from the usual goals of preventing or curing diseases, especially cancers, there are the possibilities of changing inheritance by modifying the DNA of eggs and/or sperm of organisms.

Many other uses can also already be foreseen, such as making mice, carriers of Lyme disease, immune to Lyme disease, or in wiping out populations of mosquitos that carry malaria through alteration of their ability to reproduce. But one must always be wary of, and on the lookout for, possible unintended deleterious consequences. Most likely many clever uses have not yet been conceived; the limits are, it seems, only in our imaginations. We'll just have to await the flowering that will likely emerge from humanity's collective brains. One early result: there has already been apparent success with a protocol to largely eliminate the serious symptoms of sickle cell anemia. Very encouraging.

One development shows how economically important at least some biologists think that this technique is likely to become: there is already a highly contentious lawsuit well in progress over a patent for applying this technique to humans.

The Spread of Humans and Related Organisms around the World

Humans, such as we—*Homo sapiens*—and others from the genus *Homo*, seem to have first evolved in Africa and later spread around the globe. Here I give just a few glimpses of the evidence behind this claim. The field is far too vast for me to give anything even approaching a comprehensive view.

Rather recently, the oldest known human fossil outside of Africa was discovered in Saudi Arabia: a finger. This finger has been dated via the radioactivity method, with the secure result of an age of 88,000 years. In another part of the world, Australia, what is the oldest known human remain? A fossil that weighs in at an age of about 65,000 years. These results imply that it may have taken no more than about 23,000 years for humans to have spread around the world from Arabia to Australia. Why "may"? Because oldest fossils found to date may well not represent dates of first forays of humans to those respective locales.

Older remains of humanlike organisms have been found outside Africa, for example on an Indonesian island. These, named *Homo floresiensis*, are dated as being about 100,000 years old. However, stone tools found nearby, dated to about 190,000 years ago, imply that some type of *Homo* organisms were in Indonesia far earlier than 100,000 years ago. The *Homo floresiensis* are very unusual in that, when alive, they were dwarfs, only about 106 centimeters (about 3.5 feet) in height and weighing about 27 kilograms (about 60 pounds). Do these individuals represent a new discovery of an extinct humanlike family or modern humans somehow pathologically dwarfed or . . . ? If the second, then the comments on the finger found in Saudi Arabia are null and void, except for the caveat, which would be supported.

Apparently more closely related to modern humans were Neanderthals and Denisovans. Neanderthals first appeared, we know from fossils, at least 400,000 years ago and disappeared from the world about 30,000 years ago. Studies of fragments of Neanderthal DNA found in western Europe show that Neanderthals, apparently a very talented people, mated with humans and contributed about 4 percent of our DNA, mainly affecting our metabolism and our cognitive abilities. Of Denisovans, much less is known. Among humans, the only evidence of matings with Denisovans apparently stems from studies of a tooth found in eastern Siberia, which also indicates matings with Melanesians, aboriginal Australians, and Papuans, which apparently have about 5 percent of their DNA from Denisovans. Statements about these alleged matings and percentages, I believe, are somewhat preliminary.

The latest discovery, published in April 2019, is of *Homo luzonensis*. It is based on a total of thirteen bones (seven teeth, two hand, three foot, and one thigh bone from two adults and one child) and is dated as being about 65,000 years old. These bones were found on the island of Luzon in the Philippines, which is about 640 kilometers from Asia. How did *H. luzonensis* get to this remote island so long ago?? Even a glacial period, tying up a lot of water in ice, seems not to be nearly enough to explain this migration. Stay tuned as scientists learn more about these organisms, and about the exodus from Africa and subsequent spread around the world of other *Homo* organisms and of humans.

The study of the peopling of the world, including the so-called New World, is now flourishing with the addition of the DNA tool to the armamentarium of the researchers, and we can expect to see and hear the results in the media of much more progress as time goes by.

TWENTY-TWO

The Search for Extraterrestrial Life

This last chapter should be peppered with fun, marred by only a few exceptions for which hard conceptual work is required. We are setting out to describe the search for extraterrestrial life, the stuff both of serious study and of science fiction throughout our lives. We will break down this search into three parts: elsewhere in our solar system (other planets and their satellites, if any), interstellar space (molecular precursors to life), and planets (and their satellites, if any) around other stars. We will separately discuss the search for extraterrestrial *intelligent* life (SETI), considering two compound questions: If "they" exist, will they signal us, and if so how? Will they visit us, and if so when?

Life Elsewhere in Our Solar System

What are the prospects for the existence of life elsewhere in our solar system? Let's first examine the planets. By all odds, Mercury seems inhospitable to life, either on or under the surface, or in its polar regions. The temperature seems too high and the atmosphere virtually nonexistent. Past life there, early in the solar system history, also seems unlikely—but not impossible; its past, including its location in the solar system, may have been a lot different from its present situation. What about Venus? Surface life seems unlikely; the average surface temperature there is now over 700 K, a temperature at which even tardigrades could not survive. What about subsurface life?

What about life farther up in the atmosphere, where the temperatures are moderate? We do not know. Until recently, few relevant U.S. scientists were pushing an investigation of the possibilities, so it seemed unlikely that any such search would be carried out soon, unless another country or combination took up this challenge. But that situation changed dramatically when a few years ago it was reported that phosphine, a molecule consisting of one phosphorus and three hydrogen atoms, was reported to have been detected in Venus's atmosphere. Since no one knows any nonbiological origin for this molecule, the interest in searching for life in Venus's atmosphere suddenly became urgent, even though the reliability of the detection of phosphine there has been seriously questioned. Spacecraft mission(s) may in fact soon follow unless the reported detection of phosphine is definitively shown to be spurious.

Which now brings us to Mars. Ever since the 1880s, when the Italian astronomer Giovanni Schiaparelli studied the surface of Mars with a telescope and noted the existence there of *canali*, there has been increasing interest in the possibility of life on Mars. Thus, great publicity and funds were put into this concept early in the twentieth century by Percival Lowell, a Boston Brahmin.

Skipping to the space age: our first serious mission to detect the presence of life on Mars was the Viking mission, which had two landers that set down on Mars's surface on July 4, 1976, the two-hundredth anniversary of the United States' Declaration of Independence. It was technologically sophisticated for forty-plus years ago; the results were at first considered equivocal, with a possible positive result, and then later it was definitively decided that no evidence of life had been found. Present landers roaming over the surface of Mars seem to have found evidence of some chemicals needed to construct life on earth. As a result, the publicity machines are stirring up interest in the possibility of finding evidence of past life on Mars. So far no positive results. In the future . . . ?

Beyond Mars lie the asteroids and the gaseous, cold outer planets—with at least one exception: Jupiter is known to radiate more energy than it receives from the sun. We believe that the asteroids and the minor planets, like Pluto, do not have the atmospheric raw materials that we think are needed for life, if indeed they have thick-enough atmospheres. There is here, too, at least one exception: meteorites recovered from parts of asteroids impacting on the earth have been found to contain trace amounts of amino acids. How do we know that these small amounts are not contamination from handling of the meteorites on the earth? The answer is another illustration of the unity of science, the chemistry of biology. In virtually all of the biology on earth, amino acids are found to be left-handed. In the meteorites, by contrast, the amino acids found were about half left-handed and half right-handed, demonstrating that they were not a result of contamination after the meteorite's landing on earth.

What, by the way, do we mean by left-handed and right-handed molecules? They are two forms of a molecule that are mirror images of each other and can't, by being turned in any way, be made equivalent—that is, be made indistinguishable from each other.

Despite the presence of these biological molecules, no one expects to find life on asteroids. Similarly, we do not consider the gaseous outer planets—Jupiter, Saturn, Uranus, and Neptune—to be prime targets for the search for extraterrestrial life, perhaps, though, without sufficient reason. What about more earthlike objects in the outer solar system, like Pluto? Scientists have given some thought to the possibility of life existing on such bodies; although some possible scenarios have been conceived, the general consensus is that such must be considered a real long shot.

What about the satellites of these outer planets? Offhand, one might rule them out as sources of extraterrestrial life based on their expected frigid temperatures. But that would be too quick, and, in fact, an incorrect conclusion. How is that? Some satellites of outer planets turn out to be (unexpectedly) warm below their surfaces and, hence, possible locales for life. Why would they be warmer than expected from solar heating? The answer is to be found in tidal heating. Everyone is familiar with the tides in the oceans on earth, which are due primarily to the moon and secondarily to the sun. In the cyclical rising and falling of tides, parts of the earth—land and water—rub against their neighboring parts. It is just like what happens when one rubs two sticks together: the rubbing heats up the materials due to the friction between the two as a result of the rubbing. Because of the orbital eccentricity (noncircularity) of the orbits of some of the satellites, the corresponding tides—the breathing out and in of the material as it passes closer to, and farther away from, the planet—cause the material composing the satellites to warm up. The energy for the warmth comes mostly from the orbital motions of the satellites. The energy in these motions is far, far greater than that in the tides, so there is no noticeable effect of the dissipation of heat via the tides on the orbital motions of the satellites over our lifetimes. Although this discussion is woefully incomplete, I trust that it gives at least a slight flavor (whiff) of how satellites can be warmed by tidal interactions with their planet.

Almost all scientists, in particular those involved in the Voyager flybys of Jupiter and its satellites in the late 1970s, did not realize the extent of this potential for so much warming of satellites from tides. And they were, understandably, so busy looking at the huge number of photographs taken by the first Voyager spacecraft that they missed seeing the eruption of gases from the surface of the innermost Galilean moon, our friend Io. A Jet Propulsion Laboratory technician, Linda Morabito, noticed a smudge on a photo of the surface of Io and called it to the attention of her supervisor, Stephen Synnott, who coincidentally was a former doctoral student of mine. He recognized the smudge for what it was. Thus began the study of such outpourings, which on Io were laden with sulfur gases—inimical to life as we know it. (Although not known to Linda at this time, possible volcanic activity due to tides on the inner Galilean satellites of Jupiter had been predicted in a publication by Stan Peale at UC Santa Barbara and his colleagues elsewhere, just before Voyager's encounter with Jupiter, for which they justly became famous.)

Planetary scientists now concentrate for a possible source of extraterrestrial life, however, on the second innermost of the four Galilean moons of Jupiter, Europa. It is also warmed by tides. The surface of Europa is relatively free from craters, thus pointing to its relative youth. That is, the present surface was likely the result of relatively recent melting. Also, oxygen has been detected in Europa's (thin) atmosphere. Inferences about its internal gravity field from radio tracking of spacecraft orbiting in the Jovian system are consistent with a water ocean being present somewhat below its surface. This combination of clues has led to considerable pressure from the planetary science community to mount a large spacecraft mission to Europa to

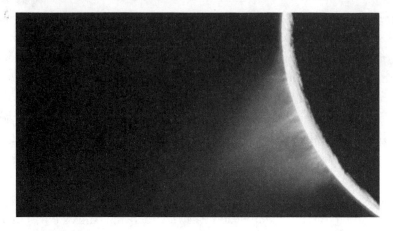

Figure 22.1. A massive outpouring from the surface of Enceladus of gases, turned to ice in the cold of space. NASA.

investigate the possibilities of life being present there. Such a mission would likely be in the multibillion-dollar class. Wait and see whether or when results are obtained; I am very unlikely to be around, although I would love to be!

Could any other moons in our solar system reasonably be inferred to harbor life? So far, the main other contender is Enceladus, one of Saturn's moons. The Cassini spacecraft captured in action volcanic outgassing from Enceladus's surface (fig. 22.1). This emerging material is likely ice; its source could be a liquid ocean underneath the surface. The case is not clear, despite analysis of spacecraft tracking data, which yields estimates of the detailed gravity field of Enceladus consistent with liquid water of a considerable thickness. Estimates of the energy involved in this outgassing are sufficiently great as to indicate that it may be only a sporadic phenomenon. Time will tell, as they say.

Another entry into the solar-system-life sweepstakes is Saturn's largest satellite, Titan. It, too, seems to have liquid below, and perhaps on, its surface. Recently, the molecule vinyl cyanide has been identified in Titan's atmosphere. Why should we care? Because some biology experts believe that this molecule could be critical to the formation of cell membranes, a key ingredient to life as we know it.

To summarize for the solar system: life *may* exist beyond the earth. Satellites of planets *may* be the most likely habitats. Sophisticated flagship (multibillion-dollar) missions *may* soon be undertaken to search for life on such satellites. Present and future missions will definitely search for evidence of present and past life on Mars, and, in the near future, probably for evidence of life in Venus's atmosphere. Come what may, though, analogs of humans are *extremely* unlikely to be uncovered elsewhere in our solar system.

Astrochemistry

We now expand our horizons beyond the solar system. What may the rest of the universe have to offer in the way of life? To start on a positive note, this field of science has been dubbed "astrobiology." It is, however, the only scientific field that I know with no data! I am being only half facetious—the biology half.

We begin with astrochemistry. This subfield asks, for example, what atoms and molecules are present in outer space. How can we find out what's there? We look primarily with radio telescopes. Why radio? Answer: space is mostly very cold. So what? As we learned in the first section of this book, the peak of cold, black-body radiation is in the radio part of the spectrum. Hence, we look in that part of the spectrum for radiation from cold material. Hydrogen is the most common element, by far, in the universe. It was natural to look first for it.

Not until World War II were quantum-mechanical calculations performed, in war-plagued Holland, that told us where in the radio spectrum to look for evidence of hydrogen in space. These calculations were carried out by the physicist Hendrik van de Hulst at the suggestion of his compatriot the Dutch astronomer Jan Oort. But it was not until 1951, at Harvard, that Ed Purcell and "Doc" Ewen succeeded in detecting signals from hydrogen in outer space, using a receiving horn placed outside the window on the south side of Harvard's physics laboratory's fourth floor, and a then sophisticated receiver system. They beat a group in Holland and one in Australia, but Purcell, being the gracious person he was, waited until the other groups had also detected hydrogen in space so that all three groups could publish in the same issue of the journal *Nature*.

The next detection was made of the hydroxyl radical OH at MIT in 1963, largely by Sander Weinreb, then a graduate student working on this project for his doctoral thesis. Five years later, water and ammonia were detected at UC Berkeley by Charles Townes and his research group. The pace soon picked up. Most molecules were found in star-forming regions. The basis for their detections was their unique spectral signatures. Since it is still beyond our capability to calculate for molecules the precise spectral values of their radiations from first principles, one needs to make laboratory measurements to predict or to verify these values; either can come first. Some spectral lines have been detected in outer space that are still unidentified. Over 20 percent of the lines detected in outer space, and identified, were first done at the then Harvard-Smithsonian Center for Astrophysics, through a program developed by Patrick Thaddeus, and later taken over by Mike McCarthy, of combined observations and clever laboratory measurements of spectral lines of many molecules.

The list of molecules identified in space so far contains more than two hundred entries (table 22.1). One of the molecules recently discovered in space has sixty carbon atoms and is shaped like a soccer ball, with five- and six-sided figures on the outside. It is named buckminsterfullerene, after its discoverer on earth, Buckminster Fuller; in slang, these molecules are called buckyballs. One point is paramount, the domination of carbon-based molecules. Is this a real effect, somehow a bias resulting from the observations, or a bit of both? Not yet clear.

Extrasolar Planets

For decades, if not centuries, people have wondered about the existence of planets around other stars (extrasolar planets). How could we find out whether any exist? Why can't we just use big telescopes and look? We have two strikes against us: the planet's dimness and the blinding light from its star, which would, from earth, in angle be very close to the planet. Indirect means of detection therefore came to the fore. We now know, and have

Table 22.1 Summary of molecules discovered
in outer space through March 2016

Number of atoms in molecule	Number of molecules discovered
2	43 (many with carbon)
3	43 (many with carbon)
4	27
5	19
6	16 (carbon dominance)
7	10 (carbon dominance)
8	11 (carbon dominance)
9	10 (carbon dominance)
>10	15 (carbon dominance)

Total = 194

used, five such means: radial velocity (Doppler shift or wobble), transit, pulsar timing, gravitational-lensing, and astrometry.

Here I discuss only the first two methods in detail, briefly describe the third, and skip the last two altogether, which I mention just to let you know that others exist. The radial-velocity method is based on a principle: when two objects orbit each other in space, each of the two objects moves around their (common) center of mass. What, you might wonder, defines the center of mass? The center of mass of two mass points, for example, is defined such that it lies on the line joining the two mass points at the place where the product of each mass and its distance from this center of mass is identical. Thus, as the planet orbits the star, the star orbits the planet.

In the radial-velocity (component of velocity along our line of sight to the star) method, one observes the radial velocity of the star to see if it has a cyclic motion, which would represent the star moving around its mutual center of mass with the planet. The period of this cyclic radial-velocity signal would be equal to the orbital period of the planet about the star. The amplitude of this periodic variation of the radial velocity would appear to us to be its greatest were the earth to lie in the orbital plane of this mutual star-planet motion. If the line of sight to the earth were perpendicular to this orbital plane, then there would not be any variation in radial velocity that we could detect, and we would not know via such measurements that there was a planet in orbit about the star. What can one learn about the planet from such a varying radial-velocity curve? Most important, we can deduce a lower bound on the mass of the planet; the details of this deduction are, however, beyond the scope of this book.

The second method of detecting extrasolar planets is called the transit method. This method requires the planet to pass in front of the star as seen from the earth—a transit. When the planet so passes, it will block out some of the light from the star that would otherwise have reached the earth. By monitoring the intensity of the light from the star, we can detect the passage of a planet in front of it by the (small) diminution of the intensity of the light that accompanies this passage (fig. 22.2). By knowing the percentage of this reduction in brightness and estimating the star's size from a combination of observations and stellar models, one can infer the size of the planet (can you do this calculation?).

We can compare the strengths and weaknesses of these two methods: the radial-velocity method favors massive, close-in planets with the line of

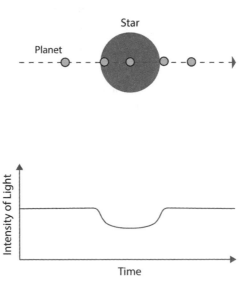

Figure 22.2. Changes in the apparent brightness (in figure: intensity of light) of a star as a planet transits in front of the star. Courtesy of David Shapiro.

sight from the earth in or near the plane of the planet's orbit; with suitable subsidiary measurements and calculations, the *mass*, but *not* the *size*, of the planet can be determined. The transit method favors large-sized, and is even more biased toward, close-in planets, with the line of sight from the earth in or near the plane of the planet's orbit, as well. Estimates of the *size*, *not* the *mass*, of the planet can be determined from the transit and suitable other observations and models related to the star, whose details I omit.

How has the search for extrasolar planets been progressing? Let's review. In 1989 there was a discovery at the Center for Astrophysics by Dave Latham, Tsevi Piran, and their colleagues of a possible planet around a sunlike star, via the radial-velocity method. Why "possible"? Because of the unknown angle between the line of sight from the earth and the orbital plane of the putative planet, one could not be sure whether this angle was moderate in size or whether it was nearly 90 degrees. In the latter case, the mass of the putative planet might have been large enough to be, for example, a brown-dwarf star rather than a true planet. That is, it could have been—or be—a double-star system rather than a planet-star system.

In 1992 there was the startling discovery, by Aleksander Wolszczan and Dale Frail, of two planets orbiting around a *pulsar*. A pulsar is an extremely dense, extremely small star, about 10 kilometers in radius, that rotates very rapidly—some pulsars even rotate many times per second (up to the order of a *thousand or more times a second*)—and gives off radiation, mostly in the radio part of the spectrum, in a particular direction from the pulsar's perspective; if the geometry is right, this radiation arrives at the earth in pulses at extremely regular intervals, set by the rotation rate of the pulsar. These intervals are, of course, modified by the motion of the pulsar with respect to us. The variations of the times of arrival at the earth of the pulses transmitted by this pulsar with two orbiting planets indicated that the pulsar was orbiting about the center of mass of this three-body system, the pulsar and its two planets. Because this was such an unexpected situation to find, many people viewed it somewhat skeptically. However, when the gravitational effects of each planet on the other could be seen clearly on the corresponding changes in the times of arrival of the pulses at the earth, all skepticism vanished. No other such pulsar-planet system has yet been definitively discovered.

Three years on, in 1995, came the unequivocal discovery of the first planet around a sunlike star, from radial-velocity observations, as in 1989, but this time by Michel Mayor and Didier Queloz, for which they received the 2019 Nobel Prize in Physics. About an additional four years later, in 1999, came the detection of the first planet around a sunlike star via the transit method.

The dam thus started breaking in 1995. By now, more than five thousand extrasolar planets are known, thanks mainly to the Kepler satellite, which used the transit method to discover planets around other stars. Kepler, launched in March 2009, has a 1-meter-diameter reflecting telescope with a field of view of 100 square degrees. (A square degree on the sky can be thought of as a patch, one degree on a side, analogous to a square unit of area on a flat surface.) Kepler stared continuously at about 150,000 stars from an earth-trailing orbit whose period is about 373 days. Because of a later failure in part of the spacecraft equipment that controls the pointing, Kepler's mission was changed and it then surveyed a larger part of the sky, sweeping in a band about the orbital plane of the spacecraft. The name of its new mission was K2, for obvious reasons; this mission is now also over.

What about further space missions? One launched in 2018, called TESS (Transiting Exoplanet Survey Satellite), is now monitoring millions of stars spread over the entire sky. TESS has already gathered data on more than five thousand transitlike signals, most of which are likely from transiting planets; for most of them, follow-up observations to confirm or deny need to be carried out. This total includes hundreds of earth-sized planets. Why do I point out the number of earth-sized planets? We are, after all, seeking evidence for extraterrestrial life. We think, based on our only experience, that we are most likely to find it in habitats like our own—perhaps a very narrow, parochial viewpoint that the future may expose as unduly restrictive. We are, somewhat contradictorily, also not being so restrictive in our own solar system by preparing, for example, to search for life on Europa and/or Enceladus and/or Titan.

What have we learned so far from our discoveries of extrasolar planets? We have data, which have allowed us to produce estimates of masses, radii (and hence densities) of many extrasolar planets, and some information on atmospheres, far from all information of relevance for each in regard to the presence or absence of life there. But new information flows in daily; some of it is startling, quite different from prior expectations. It is hard to keep up, as it is a flourishing new field.

What are some of the biggest surprises? Many of these planets are Jupiter sized and larger. Moreover, a large number have remarkably short orbital periods, some of only a very few days and one of a mere 4.2 *hours* (equivalent to a slightly over four-hour *year*). Thus, in contrast to our solar system, many others have planets orbiting extremely close to their star, as follows from their short orbital periods and Kepler's third law. Some of these planets, too, move in very eccentric orbits, again unlike any planets in our solar system. The planet with the highest orbital eccentricity yet seen has an orbital eccentricity of 0.97, nearly at the limit of 1. These results are clearly biased by our means of observation. For example, with the two methods discussed in detail, we are much more likely to discover planets with very short orbital periods than those with very long orbital periods. Can you think why?

The Search for Extraterrestrial Life

What about extraterrestrial life? There are no detections yet. If there were, you undoubtedly would have heard! What do we look for? As noted above, we look for earthlike planets. Not only do we look for planets of earth size, but we look for them to be orbiting in the so-called habitable zone of their star. Basically, a habitable zone is the region near a star in which liquid water can exist on the surface of an earthlike planet. This requirement is due to our first guess that liquid water is necessary for life. We also look for planets that have atmospheric spectral signatures indicative of molecules that we associate with life, such as oxygen—again, an anthropomorphic approach.

Note, also again, that our search for extraterrestrial life in our solar system violates these principles: Europa, Enceladus, and Titan all lie way outside of the habitable zone of our star. What about satellites of planets around other stars? None has yet been detected, but not for lack of trying. Such detections are technically very challenging, but my guess is that success will be at hand within the next decade. I find it hard to believe that conditions and processes, which led to the formation of satellites in our solar system, would not be present and in operation around other stars as well. When these distant satellite systems are discovered, perhaps with more sensitive instruments, the search space for habitable worlds will perforce vastly expand.

Keep in mind: the discovery of extraterrestrial life can be made only once in the history of humankind. Your generation may be the one to accomplish this truly historic feat!

Search for Extraterrestrial Intelligence

We move to our next big—and last!—question: Is there extraterrestrial *intelligence* elsewhere in the universe and, if so, would "they" try to communicate with and/or visit us? Back in the late 1950s, Giuseppe Cocconi and Philip Morrison at Cornell University suggested looking for signals transmitted to us by others, using the frequency of the main spectral line of hydrogen. That suggestion was picked up by radio astronomers, and the equipment devoted to this search has increased in sophistication and sensitivity in the interim, including a "SETI at home" project (SETI stands for search for extraterrestrial intelligence). The largest such effort is now being conducted in Northern California with the Allen array of forty-two radio telescopes. So far there have been no detections of signals known, or even suspected, to have been sent by extraterrestrial intelligent beings. This effort will very soon be greatly outclassed by a new initiative, named Breakthrough Listen.

In the 1970s, Charles Townes at UC Berkeley suggested looking for narrow-band laser signals from extraterrestrial intelligence. As a result of this suggestion, we had a 1.8-meter-diameter optical telescope in Harvard, Massachusetts, run by Paul Horowitz of Harvard University's physics department, observing the sky for signs of such signals, using very sophisticated data processing algorithms. So far, no success—or you would have heard, as for any success in the search for extraterrestrial life.

Where are the extraterrestrials? This is in fact a deep question. Why have no extraterrestrials visited us, either in person or far more easily via radio signals? This question was popularized after Enrico Fermi, the famous Italian American physicist, apparently casually posed it at a lunch in 1950. It has, after elaboration, become known as the Fermi paradox. In 2017, an object from beyond the solar system passed near the earth and then left the solar system. From the apparent influence of sunlight pressure on its orbit, scientists deduced that it may be extraordinarily thin and thus possibly created elsewhere by intelligent life. This possibility, promulgated and amplified by my Harvard colleague Avi Loeb, led to the story about it going viral, worldwide. This story is, however, only a possibility; we do not have nearly sufficient data to conclude that this object was in fact made by other intelligent creatures. But we should be alert in the future to such objects and make sure we are prepared and observe them in as detailed a manner as technology will then allow, perhaps even guiding one to earth while being appropriately careful!

Some of the possible reasons we haven't been visited or contacted via, say, radio signals by extraterrestrials are: they don't exist; they haven't yet (accounting for the travel time of beings or of light) developed the needed technology; they are not interested in communicating with, or in traveling to, us; or they die out, or become incapable of communication, too fast; hence, there is no overlap with us at present (again with light travel time having been accounted for).

Suppose sometime in the future we reliably detect signals from extraterrestrial intelligent life. What then? It would be enormously exciting. But we would likely have a problem: it is likely—however we wish to define "likely"—that such signals will originate at least hundreds of light-years away from us. Not only will we have the problem of understanding each other's languages, or means of communication, but we would have to wait at least a round-trip light travel time to receive any response to messages we transmit. The first such transmissions might involve universals, such as the value of pi, the ratio of the circumference to the diameter of a circle; how, though, do we send these "unambiguously"? Think about possible means! At present levels of human longevity, any conversation would probably have to be a multigenerational affair from our point of view. Such would be rather a shocking change from our current instant gratification culture. But I think it would be well worth it and indeed irresistible. Do you?

On that note, we end our short survey of science and, concomitantly, this book. I hope that you have found it both fun and enlightening!

Alberts, Bruce, Dennis Bray, Julian Lewis, Martin Raff, Keith Roberts, James D. Watson, Nigel Orme, and Kay Hesketh-Moore. *Molecular Biology of the Cell.* 3rd ed. New York: Garland, 1994.

Alexander, Robert McNeill. "Estimates of the Speeds of Dinosaurs." *Nature* 261 (1976): 129–130.

Alvarez, Walter. *A Most Improbable Journey: A Big History of Our Planet.* New York: W. W. Norton, 2017.

———. *T. rex and the Crater of Doom.* Princeton, N.J.: Princeton University Press, 1997.

Amend, Jan P., and Everett L. Shock. "Energetics of Amino Acid Synthesis in Hydrothermal Ecosystems." *Science* 281 (1998): 1659–1662.

Amiot, Romain, Christophe Lécuyer, Eric Buffetaut, and Gilles Escarguel. "Oxygen Isotopes from Biogenic Apatites Suggest Widespread Endothermy in Cretaceous Dinosaurs." *Earth and Planetary Science Letters* 246 (2006): 41–54.

Amiot, Romain, Xu Wang, Zhonghe Zhu, Xiaolin Wang, Eric Buffetaut, Christophe Lécuyer, Zhongli Ding, et al. "Oxygen Isotopes of East Asian Dinosaurs Reveal Exceptionally Cold Early Cretaceous Climates." *Proceedings of the National Academy of Sciences* 108 (2011): 5179–5183.

Avery, Oswald T., Colin M. MacLeod, and Maclyn McCarty. "Studies on the Chemical Nature of the Substance Inducing Transformation of Pneumococcal Types: Induction of Transformation by a Desoxyribonucleic Acid Fraction Isolated from Pneumococcal Type III." *Journal of Experimental Medicine* 79 (1944): 137–158.

Barnhart, Edwin L. "Reconstructing the Heavens: Archaeoastronomy and the Ancient Maya World." *Mercury* (January–February 2004): [22]–29.

Barondes, Samuel H., and Marshall W. Nirenberg. "Fate of a Synthetic Polynucleotide Directing Cell-Free Protein Synthesis II. Association with Ribosomes." *Science* 138 (1962): 813–817.

Baucon, Andrea. "Da Vinci's Paleodictyon: The Fractal Beauty of Traces." *Acta Geologica Polonica* 60 (2010): 3–17.

———. "Italy, the Cradle of Ichnology: The Legacy of Aldrovandi and Leonardo." *Studi Trentini di Scienze Naturali, Acta Geologica* 83 (2008): 15–29.

———. "Leonardo da Vinci, the Founding Father of Ichnology." *Palaios* 25 (2010): 361–367.

Beaudry, Amber A., and Gerald F. Joyce. "Directed Evolution of an RNA Enzyme." *Science* 257 (1992): 635–641.

Becker, George F. "Halley on the Age of the Ocean." *Science* 31 (1910): 459–461.

Bedini, Silvio A. *The Pulse of Time: Galileo Galilei, the Determination of Longitude, and the Pendulum Clock.* [Florence]: L. Olschki, 1991.

Benton, Michael J. "Dinosaurs." *Current Biology* 19 (2009): R318–R323.

———. "Fossil Quality and Naming Dinosaurs." *Biology Letters* 4 (2008): 729–732.

———. "Fossil Record: Quality." In *Encyclopedia of Life Sciences.* New York: John Wiley and Sons, 2005.

———. "How to Find a Dinosaur, and the Role of Synonymy in Biodiversity Studies." *Paleobiology* 34 (2008): 516–533.

———. *Introduction to Paleobiology and the Fossil Record.* Hoboken, N.J.: Wiley-Blackwell, 2009.

———. "Naming Dinosaur Species: The Performance of Prolific Authors." *Journal of Vertebrate Paleontology* 30 (2010): 1478–1485.

———. "Phylocode: Beating a Dead Horse?" *Acta Palaeontologica Polonica* 52 (2007): 651–655.

———. "The Red Queen and the Court Jester: Species Diversity and the Role of Biotic and Abiotic Factors through Time." *Science* 323 (2009): 728–732.

———. "Studying Function and Behavior in the Fossil Record." *PLoS Biology* 8 (2010): e1000321.

Benton, Michael J., and Philip C. J. Donoghue. "Paleontological Evidence to Date the Tree of Life." *Molecular Biology and Evolution* 24 (2007): 26–53.

Bernstein, Max P., Jason P. Dworkin, Scott A. Sandford, George W. Cooper, and Louis J. Allamandola. "Racemic Amino Acids from the Ultraviolet Photolysis of Interstellar Ice Age Analogues." *Nature* 416 (2002): 401–403.

Binzel, Richard P., Alessandro Morbidelli, Sihane Merouane, Francesca E. DeMeo, Mirel Birlan, Pierre Vernazza, Cristina A. Thomas, Andrew S. Rivkin, Schelte J. Bus, and Alan T. Tokunaga. "Earth Encounters as the Origin of Fresh Surfaces on Near-Earth Asteroids." *Nature* 463 (2010): 331–334.

Blundell, Derek J., and Andrew C. Scott, eds. *Lyell: The Past Is Key to the Present.* London: Geological Society, 1998.

Bobis, Laurence, and James Lequeux. "Cassini, Rømer, and the Velocity of Light." *Journal of Astronomical History and Heritage* 11 (2008): 97–105.

Bochkarev, Nikolai G., Eugenia A. Karitskaia, and Nikolai I. Shakura. "Calculation of the Ellipsoidality Effect in Close Binaries with a Single Optical Component." *Soviet Astronomy* 23 (1979): 8–16.

Bohr, Niels. "Atomic Models and X-Ray Spectra." *Nature* 92 (1914): 553–554.

———. "On the Constitution of Atoms and Molecules I." *Philosophical Magazine* 26 (1913): 1–25.

———. "On the Constitution of Atoms and Molecules II." *Philosophical Magazine* 26 (1913): 476–502.

———. "On the Constitution of Atoms and Molecules III." *Philosophical Magazine* 26 (1913): 857–875.

Boorstin, Daniel J. *The Discoverers*. New York: Random House, 1983.

Bowditch, Nathaniel. "An Estimate of the Weight, Direction, Velocity and Magnitude of the Meteor That Exploded over Weston in Connecticut, December 14, 1807. With Methods of Calculating Observations Made on Such Bodies." *Memoirs of the American Academy of Arts and Sciences* 3 (1815): 213–236.

Boyko, Adam R., Pascale Quignon, Lin Li, Jeffrey J. Schoenebeck, Jeremiah D. Degenhardt, Kirk E. Lohmueller, Keyan Zhao, et al. "Simple Genetic Architecture Underlies Morphological Variation in Dogs." *PLoS Biology* 2010, https://doi.org/10.1371/journal.pbio.1000451.

Brasier, Martin D., Owen R. Green, Andrew P. Jephcoat, Annette K. Kleppe, Martin J. Van Kranedonk, John F. Lindsay, Andrew Steele, and Nathalie V. Grassineau. "Questioning the Evidence for Earth's Oldest Fossils." *Nature* 416 (2002): 76–81.

Brenner, Sydney. "New Directions in Molecular Biology." *Nature* 248 (1974): 785–787.

Briggs, Derek E. G. "The Role of Decay and Mineralization in the Preservation of Soft-Bodied Fossils." *Annual Review of Earth and Planetary Sciences* 31 (2003): 275–301.

Brown, Guy C. "NO Says Yes to Mitochondria." *Science* 299 (2003): 938–939.

Browne, Janet. *Charles Darwin*. Vol. 2: *Power of Place*. Princeton, N.J.: Princeton University Press, 2002.

Bryant, J. Daniel, and Philip N. Froelich. "A Model of Oxygen Isotope Fractionation in Body Water of Large Mammals." *Geochimica et Cosmochimica Acta* 59 (1995): 4523–4537.

Bryson, Bill. *A Short History of Nearly Everything*. New York: Broadway Books, 2003.

Buffetaut, Eric, David Martill, and François Escuillé. "Pterosaurs as Part of a Spinosaur Diet." *Nature* 430 (2004): 33.

Burchfield, Joe D. "Darwin and the Dilemma of Geological Time." *Isis* 65 (1974): 301–321.

———. *Lord Kelvin and the Age of the Earth*. New York: Science History Publications, 1975.

Burney, William. "Marine Chair." In *Falconer's New Dictionary of the Marine: 1815 Edition*, by William Falconer, edited by William Burney. London: Chatham, 2006.

Capra, Fritjof. *The Science of Leonardo: Inside the Mind of the Great Genius of the Renaissance.* New York: Doubleday, 2007.

Carrano, Matthew T. "Body-Size Evolution in the Dinosauria." In *Amniote Paleobiology: Perspectives on the Evolution of Mammals, Birds, and Reptiles; a Volume Honoring James Allen Hopson,* 225–268. Edited by Matthew T. Carrano et al. Chicago: University of Chicago Press, 2006.

Cassini, Jean Dominique. "Monsieur Cassini and His New and Exact Tables for the Eclipses of the First Satellite of Jupiter, Reduced to Julian Stile, and Meridian of London." *Philosophical Transactions* 18 (1694): 237–256.

Casson, Lionel. *Travel in the Ancient World.* London: Allen and Unwin, 1974.

Chakrabarti, Sonali, and Sandeep K. Chakrabarti. "Can DNA Bases Be Produced during Molecular Cloud Collapse?" *Astronomy and Astrophysics* 354 (2000): L6–L8.

Champoux, James J., and Renato Dulbecco. "An Activity from Mammalian Cells That Untwists Superhelical DNA: A Possible Swivel for DNA Replication (Polyoma/Ethidium Bromide/Mouse-Embryo Cells/Dye Binding Assay)." *Proceedings of the National Academy of Sciences* 69 (1972): 143–146.

Chargaff, Erwin. "Building the Tower of Babble." *Nature* 248 (1974): 776–779.

———. "Chemical Specificity of Nucleic Acids and Mechanism of Their Enzymatic Degradation." *Experientia* 6 (1950): 201–209.

———. "In Dispraise of Reductionism." *BioScience* 47 (1997): 795–797.

———. "Preface to a Grammar of Biology: A Hundred Years of Nucleic Acid Research." *Science* 172 (1971): 637–642.

Chiappe, Luis M., Laura Codorniú, Gerald Grellet-Tinner, and David Rivarola. "Argentinian Unhatched Pterosaur Fossil." *Nature* 432 (2004): 571–572.

Clayden, Jonathan. *Organic Chemistry.* Reprint ed. Oxford: Oxford University Press, 2001.

Cleland, Timothy P., Kristyn Voegele, and Mary H. Schweitzer. "Empirical Evaluation of Bone Extraction Protocols." *PLoS ONE* 7 (2012): e31443.

Clementz, Mark T. "New Insight from Old Bones: Stable Isotope Analysis of Fossil Mammals." *Journal of Mammalogy* 93 (2012): 368–380.

Clementz, Mark T., Anjali Goswami, Philip D. Gingerich, and Paul L. Koch. "Isotopic Records from Early Whales and Sea Cows: Contrasting Patterns of Ecological Transition." *Journal of Vertebrate Paleontology* 26 (2006): 355–370.

Cloern, James E., Elizabeth Canuel, and David Harris. "Stable Carbon and Nitrogen Isotope Composition of Aquatic and Terrestrial Plants of the San Francisco Bay Estuarine System." *Limnology and Oceanography* 47 (2002): 713–729.

Cobb, Matthew. *Life's Greatest Secret: The Race to Crack the Genetic Code.* New York: Basic Books, 2015.

Cocconi, Giuseppe, and Philip Morrison. "Searching for Interstellar Communications." *Nature* 185 (1959): 844–846.

Cohen, Jack S., and Franklin H. Portugal. "The Search for the Chemical Structure of DNA." *Connecticut Medicine* 38 (1974): 551–557.

Colbert, Edwin H. "Feeding Strategies and Metabolism in Elephants and Sauropod Dinosaurs." *American Journal of Science* 203A (1993): 1–19.

Colson, Francis Henry. *The Week: An Essay on the Origin and Development of the Seven-Day Cycle*. Westport, Conn.: Greenwood, 1974.

Courtillot, Vincent. *Evolutionary Catastrophes: The Science of Mass Extinction*. New York: Cambridge University Press, 1999.

Crick, Francis. "The Double Helix: A Personal View." *Nature* 248 (1974): 766–769.

———. *What Mad Pursuit: A Personal View of Scientific Discovery*. New York: Basic Books, 1988.

Curry, Gordon B. "Molecular Palaeontology." In *Palaeobiology: A Synthesis*, edited by Derek G. Briggs and Peter R. Crowther, 95–100. Oxford: Blackwell Scientific, 1990.

———. "Molecular Palaeontology: New Life for Old Molecules." *Trends in Ecology and Evolution* 2 (1987): 161–165.

Curry Rogers, Kristina. *Sauropods: Evolution and Paleobiology*. Berkeley: University of California Press, 2005.

Dahm, Ralf. "The First Discovery of DNA." *American Scientist* 96 (2008): 320–327.

———. "Friedrich Miescher and the Discovery of DNA." *Developmental Biology* 278 (2005): 274–288.

———. "From Discovering to Understanding." *EMBO Reports* 11 (2010): 153–160.

Dalrymple, G. Brent. *The Age of the Earth*. Stanford, Calif.: Stanford University Press, 1991.

Darnell, J. E., Jr. "The Origin of mRNA and the Structure of the Mammalian Chromosome." *Harvey Lectures* 69 (1973–1974): 1–47.

Darwin, Charles. *From So Simple a Beginning: The Four Great Books of Charles Darwin*. Edited, with introductions, by Edward Osborne Wilson. New York: W. W. Norton, 2006.

———. *On the Origin of Species* [1859]. Reprint of 1st ed. with an introduction by Ernst Mayr. Cambridge, Mass.: Harvard University Press, 1975.

———. *On the Origin of Species by Means of Natural Selection; or, The Preservation of Favoured Races in the Struggle for Life* [1872]. Facsimile reprint of 6th ed. N.p.: Elibron Classics, 2005.

Davies, Kevin. *Cracking the Genome: Inside the Race to Unlock Human DNA*. New York: Free Press, 2001.

Davison, Charles. *Founders of Seismology*. New York: Arno Press, 1978.

Dawkins, Richard, and Yan Wong. *Ancestor's Tale: A Pilgrimage to the Dawn of Evolution*. Boston: Houghton Mifflin, 2004.

"A Demonstration concerning the Motion of Light; Communicated from Paris, in the *Journal des Scavans*, and Here Made English." *Philosophical Transactions* 12 (1676): 893–894.

Deng, Tao, Xiaoming Wang, Mikael Fortelius, Qiang Li, Yang Wang, Zhijie J. Tseng, Gary T. Takeuchi, Joel E. Saylor, Laura K. Säilä, and Guangpu Xie. "Out of Tibet:

Pliocene Wooly Rhino Suggests High-Plateau Origin of Ice Age Megaherbivores." *Science* 333 (2011): 1285–1288.

Denton, Francis M. "Einstein's Theory." *Times* (London), November 14, 1919.

Di Giulio, Massimo, and Mario Medugno. "Physiochemical Optimization in the Genetic Code Origin as the Number of Codified Amino Acids Increases." *Journal of Molecular Evolution* 49 (1999): 1–10.

Dietz, Robert S. "Continent and Ocean Basin Evolution by Spreading of Sea Floor." *Nature* 190 (1961): 854–857.

Donoghue, Philip C. J., and Jonathan B. Antcliffe. "Origins of Multicellularity." *Nature* 466 (2010): 41–42.

Donoghue, Philip C. J., and Michael J. Benton. "Rocks and Clocks: Calibrating the Tree of Life Using Fossils and Molecules." *Trends in Ecology and Evolution* 22 (2007): 424–431.

Doppler, Christian. "On the Coloured Light of Double Stars and Certain Other Stars of the Heavens: An Attempt at a General Theory Which Incorporates Bradley's Theorem of Aberration as an Integral Part." In *The Search for Christian Doppler*, by Alec Eden, 101–133. New York: Springer-Verlag, 1992.

Doudna, Jennifer, and Samuel H. Sternberg. *A Crack in Creation: Gene Editing and the Unthinkable Power to Control Evolution.* Boston: Houghton Mifflin Harcourt, 2017.

Drake, Stillman. *Discoveries and Opinions of Galileo.* Garden City, N.Y.: Doubleday, 1957.

Dutka, Jacques. "Eratosthenes' Measurement of the Earth Reconsidered." *Archive for History of Exact Sciences* 26 (1993): 55–66.

Du Toit, Alexander Logie. "Tertiary Mammals and Continental Drift: A Rejoinder to George G. Simpson." *American Journal of Science* 242 (1944): 145–163.

Eagle, Robert A., Edwin A. Schauble, Aradhna K. Tripati, Thomas Tütken, Richard C. Hulbert, and John M. Eiler. "Body Temperatures of Modern and Extinct Vertebrates from $^{13}C–^{18}O$ Bond Abundances in Bioapatite." *Proceedings of the National Academy of Sciences* 107 (2010): 10377–10382.

Eagle, Robert A., Thomas Tütken, Taylor S. Martin, Aradhna K. Tripati, Henry C. Fricke, Melissa Connely, Richard I. Cifelli, and John M. Eiler. "Dinosaur Body Temperatures Determined by Isotopic ($^{13}C–^{18}O$) Ordering in Fossil Biominerals." *Science* 333 (2011): 443–445.

Edwards, Anthony William Fairbank. "Are Mendel's Results Really Too Close?" *Biological Reviews* 61 (1986): 295–312.

Ehret, Charles F. "Organelle Systems and Biological Organization: Structural and Developmental Evidence Leads to a New Look at Our Concepts of Biological Organization." *Science* 132 (1960): 115–123.

Einstein, Albert. *The Collected Papers of Albert Einstein.* Vol. 6: *The Berlin Years: Writings, 1914–1917.* Edited by A. J. Knox, Martin J. Klein, and Robert Schulmann. Princeton, N.J.: Princeton University Press, 1996.

————. "On the Theory of the Static Gravitational Field," March 23, 1912, and "Note Added in Proof." In *The Collected Papers of Albert Einstein*, Vol. 4: *The Swiss Years: Writings, 1912–1914*, 107–120. Translated by Anna Beck. Princeton, N.J.: Princeton University Press, 1996.

El Albani, Abderazzak, Stefan Bengston, Donald E. Canfield, Andrey Bekker, Roberto Macchiarelli, Arnaud Mazurier, Emma U. Hammerlund, et al. "Large Colonial Organisms with Coordinated Growth in Oxygenated Environments 2.1 Gyr Ago." *Nature* 466 (2010): 100–104.

Erickson, Gregory M., Peter J. Makovicky, Philip J. Currie, Mark A. Norell, Scott A. Yerby, and Christopher A. Brochu. "Gigantism and Comparative Life-History Parameters of Tyrannosaurid Dinosaurs." *Nature* 430 (2004): 772–775.

Erwin, Douglas H. *Extinction: How Life on Earth Ended 250 Million Years Ago*. Princeton, N.J.: Princeton University Press, 2006.

Evans, James. *The History and Practice of Ancient Astronomy*. New York: Oxford University Press, 1998.

Eve, Arthur Stewart. *Rutherford: Being the Life and Letters of the Rt. Hon. Lord Rutherford, O.M.* New York: Macmillan, 1939.

Fahie, John Joseph. *Galileo, His Life and Work* [1903]. Dubuque, Iowa: W. C. Brown Reprint Library, [1972].

Fairbrother, Trevor J. *Leonardo Lives: The Codex Leicester and Leonardo da Vinci's Legacy of Art and Science*. Seattle: Seattle Art Museum in association with University of Washington Press, 1997.

Farquhar, James, Huiming Bao, and Mark Thiemens. "Atmospheric Influence of Earth's Earliest Sulfur Cycle." *Science* 289 (2000): 756–758.

Farlow, James O., and Michael K. Brett-Surman, eds. *The Complete Dinosaur*. Bloomington: Indiana University Press, 1997.

Felsenstein, Joseph. "Cases in Which Parsimony or Compatibility Methods Will Be Positively Misleading." *Systematic Zoology* 27 (1978): 401–410.

————. "Confidence Limits on Phylogenies: An Approach Using the Bootstrap." *Evolution* 39 (1985): 783–791.

————. "Evolutionary Trees from DNA Sequences." *Journal of Molecular Evolution* 17 (1981): 368–376.

————. *Inferring Phylogenies*. Sunderland, Mass.: Sinauer Associates, 2004.

————. "Parsimony in Systematics: Biological and Statistical Issues." *Annual Review of Ecology and Systematics* 14 (1983): 313–333.

Fischer, Irene. "Another Look at Eratosthenes' and Posidonius' Determinations of the Earth's Circumference." *Quarterly Journal of the Royal Astronomical Society* 16 (1975): 152–167.

Fisher, Ronald A. "Has Mendel's Work Been Rediscovered?" *Annals of Science* 1 (1936): 115–137.

Forbes, Eric G., Arthur Jack Meadows, and Derek Howse. *Greenwich Observatory: The Royal Observatory at Greenwich and Herstmonceux, 1675–1975.* London: Taylor and Francis, 1975.

Forest, Felix. "Calibrating the Tree of Life: Fossils, Molecules and Evolutionary Time-scales." *Annals of Botany* 104 (2009): 789–794.

Frankel, Henry. "The Development, Reception, and Acceptance of the Vine-Matthews-Morley Hypothesis." *Historical Studies in the Physical Sciences* 13 (1982): 1–39.

Frankfort, H., H. A. Frankfort, John A. Wilson, Thorkild Jacobsen, and William A. Irwin. *The Intellectual Adventure of Ancient Man: An Essay on Speculative Thought in the Ancient Near East.* Chicago: University of Chicago Press, 1946.

Freeman, Scott, and John C. Herron. *Evolutionary Analysis.* 4th ed. Upper Saddle River, N.J.: Pearson Prentice Hall, 2007.

Futuyma, Douglas J. *Evolution.* 3rd ed. Sunderland, Mass.: Sinauer Associates, 2013.

Galilei, Galileo. *Dialogue on the Great World Systems: In the Salusbury Translation.* Revised, annotated, and with an introduction by Giorgio de Santillana. Abridged text ed. Chicago: University of Chicago Press, 1955.

Gatesy, John, and Maureen O'Leary. "Deciphering Whale Origins with Molecules and Fossils." *Trends in Ecology and Evolution* 16 (2001): 562–570.

Geikie, Archibald. *Charles Darwin as Geologist: The Rede Lecture Given at the Darwin Centennial Commemoration 24 June 1909.* Cambridge: Cambridge University Press, 1909.

Gellert, Martin, Kiyoshu Mizuuchi, Mary H. O'Dea, and Howard A. Nash. "DNA Gyrase: An Enzyme That Introduces Superhelical Turns into DNA." *Proceedings of the National Academy of Sciences* 73 (1976): 3872–3876.

Gesteland, Raymond F., Thomas R. Cech, and John F. Atkins, eds. *The RNA World: The Nature of Modern RNA Suggests a Prebiotic RNA.* 2nd ed. Cold Spring Harbor, N.Y.: Cold Spring Harbor Laboratory Press, 1999.

———. *The RNA World: The Nature of Modern RNA Suggests a Prebiotic RNA.* 3rd ed. Cold Spring Harbor, N.Y.: Cold Spring Harbor Laboratory Press, 2006.

Gingerich, Owen. *The Eye of Heaven: Ptolemy, Copernicus, Kepler.* New York: American Institute of Physics, 1993.

———. "Foreword." In *Ptolemy's Almagest,* translated and annotated by Gerald J. Toomer, vii–x. London: Duckworth, 1984.

———. "Islamic Astronomy." *Scientific American* 254 (1986): 74–83.

Gingerich, Philip D. "Land-to-Sea Transition in Early Whales: Evolution of the Eocene Archaeoceti (Cetacea) in Relation to Skeletal Proportions and Locomotion of Living Semiaquatic Mammals." *Paleobiology* 29 (2003): 429–454.

Gitschier, Jane. "The Eureka Moment: An Interview with Sir Alec Jeffreys." *PLoS Genetics* 5 (2009): e1000765.

Glen, William. *The Road to Jaramillo: Critical Years of the Revolution in Earth Science.* Stanford, Calif.: Stanford University Press, 1982.

Gohau, Gabriel. "Chapter 10: Use of Fossils." In *A History of Geology*, 125–137. New Brunswick, N.J.: Rutgers University Press, 1991.

——. "Chapter 11: Uniformitarianism versus Catastrophism." In *A History of Geology*, 139–149. New Brunswick, N.J.: Rutgers University Press, 1991.

Goldstein, Bernard R. "Eratosthenes on the 'Measurement' of the Earth." *Historia Mathematica* 11 (1984): 411–416.

Gould, Stephen Jay. *Bully for Brontosaurus: Reflections in Natural History*. New York: W. W. Norton, 1991.

——. *Leonardo's Mountain of Clams and the Diet of Worms: Essays on Natural History*. New York: Harmony Books, 1998.

Gray, Michael W. "Mitochondrial Evolution." *Cold Spring Harbor Perspectives in Biology* 4 (2012): a011403.

Gregory, T. Ryan, ed. *The Evolution of the Genome*. Burlington, Mass.: Elsevier Academic, 2005.

Grellet-Tinner, Gerald, Stephen W. Wroe, Michael B. Thompson, and Qiang Ji. "A Note on Pterosaur Nesting Behavior." *Historical Biology* 19 (2007): 273–277.

Griffith, Fred. "The Significance of Pneumococcal Types." *Journal of Hygiene* 27 (1928): 113–159.

Griffiths, Anthony J. S., Susan R. Wessler, Richard C. Lewontin, and Sean B. Carroll. *Introduction to Genetic Analysis*. 9th ed. New York: W. H. Freeman, 2008.

Gulbekian, Edward. "The Origin and Value of the Stadion Unit Used by Eratosthenes in the Third Century B.C." *Archive for History of Exact Sciences* 37 (1987): 359–363.

Gurdon, John B. "Molecular Biology in a Living Cell." *Nature* 248 (1974): 772–776.

Hall, Barry G. "Comparison of the Accuracies of Several Phylogenetic Methods Using Protein and DNA Sequences." *Molecular Biology and Evolution* 22 (2005): 792–802.

Hall, Brian K., and Benedikt Hallgrimsson. *Strickberger's Evolution: The Integration of Genes, Organisms and Populations*. 4th ed. Sudbury, Mass.: Jones and Bartlett, 2008.

Halley, Edmund. "Proposal of a Method for Finding the Longitude at Sea, within a Degree or Twenty Leagues; with an Account of the Progress He Hath Made Therein, by a Continued Series of Accurate Observations of the Moon, Taken by Himself at the Royal Observatory at Greenwich." *Philosophical Transactions* 37 (1731): 185–195.

——. "A Short Account of the Cause of the Saltness of the Ocean, and of the Several Lakes That Emit No Rivers; with a Proposal, by Help Thereof, to Discover the Age of the World." *Philosophical Transactions of the Royal Society of London* 29 (1714): 296–300.

——. *The Three Voyages of Edmund Halley in the Paramore, 1698–1701*. Edited by Norman J. W. Thrower. London: Hakluyt Society, 1981.

Hartl, Daniel L., and Bruce Cochrane. *Genetics: Analysis of Genes and Genomes*. 7th ed. Sudbury, Mass.: Jones and Bartlett, 2009.

Hazen, Robert M. *Symphony in C: Carbon and the Evolution of (Almost) Everything.* New York: W. W. Norton, 2017.

Heath, Thomas Little. *Aristarchus of Samos, the Ancient Copernicus: A History of Greek Astronomy to Aristarchus, Together with Aristarchus's Treatise on the Sizes and Distances of the Sun and Moon.* Oxford: Clarendon Press of Oxford University Press, 1913.

———. *Greek Astronomy* [1932]. Reprint ed. New York: Dover, 1991.

Henry, Allison A., and Floyd E. Romesberg. "The Evolution of DNA Polymerases with Novel Activities." *Current Opinion in Biotechnology* 16 (2005): 370–377.

Herbert, Sandra. *Charles Darwin, Geologist.* Ithaca, N.Y.: Cornell University Press, 2005.

Hess, Harry Hammond. "The History of Ocean Basins." In *Petrologic Studies: A Volume in Honor of A. E. Buddington,* edited by A. E. J. Engel, Harold L. James and B. F. Leonard, 599–620. New York: Geological Society of America, 1962.

———. "Nature of Great Oceanic Ridges." In *Preprints of the First International Oceanic Congress, New York, August 31–September 12, 1959,* 33–34.

———. "Reply." *Journal of Geographical Research* 20 (1968): 6569.

Hokkanen, Jyrki E. I. "The Size of the Largest Land Animal." *Journal of Theoretical Biology* 118 (1986): 491–499.

Holmes, Arthur. "Radioactivity and Earth Movements." *Transactions of the Geological Society of Glasgow* 18 (1929): 559–606.

Holton, Gerald. "On the Origins of the Special Theory of Relativity." *American Journal of Physics* 28 (1960): 627–636.

Hopkins, William. "On the Phenomena of Precession and Nutation, Assuming the Fluidity of the Inside of the Earth." *Philosophical Transactions of the Royal Society* 129 (1839): 381–423.

Huelsenbeck, John P., Frederik Ronquist, Rasmus Nielsen, and Jonathan P. Bollback. "Bayesian Inference of Phylogeny and Its Impact on Evolutionary Biology." *Science* 294 (2001): 2310–2314.

Hunter, Graeme K. "Phoebus Levene and the Tetranucleotide Structure of Nucleic Acids." *Ambix* 46 (1999): 73–103.

Hwang, Koo-Geun, Min Huh, Martin Lockley, and David M. Unwin. "New Pterosaur Tracks (Pteraichnidae) from the Late Cretaceous Uhangri Formation, Southwestern Korea." *Geological Magazine* 139 (2002): 421–436.

Illy, József. *Albert Meets America: How Journalists Treated Genius during Einstein's 1921 Travels.* Baltimore: Johns Hopkins University Press, 2006.

Jeffreys, Alec J. "Genetic Fingerprinting." *Nature Medicine,* 11 (2005): 1035–1039.

Jeffreys, Alec J., Victoria Wilson, and Swee Lay Thein. "Individual-Specific Fingerprints of Human DNA." *Nature* 316 (1985): 76–79.

———. "Hypervariable 'Minisatellite' Regions in Human DNA." *Nature* 314 (1985): 67–73.

Jeffreys, Alec J., Maxine J. Allen, Erika Hagelberg, and Andreas Sonnberg. "Identification of the Skeletal Remains of Josef Mengele by DNA Analysis." *Forensic Science International* 56 (1992): 65–76.

Jérôme Lalande, Diary of a Trip to England, 1763. Translated from the original manuscript by Richard Watkins. 2014. https://www.watkinsr.id.au/Lalande.pdf.

Ji, Qiang, Shu-An Ji, Yen-Nien Cheng, Hai-Lou You, Jun-Chang Lü, Yong-Qing Liu, and Chong-Xi Yuan. "Pterosaur Egg with a Leathery Shell." *Nature* 432 (2004): 572.

Jianu, Coralia-Maria, and David B. Weishampel. "The Smallest of the Largest: A New Look at Possible Dwarfing in Sauropod Dinosaurs." *Geologie en Mijnbouw* 78 (1999): 335–343.

Jim, Susan, Stanley H. Ambrose, and Richard P. Evershed. "Stable Carbon Isotopic Evidence for Differences in the Dietary Origin of Bone Cholesterol, Collagen and Apatite: Implications for Their Use in Palaeodietary Reconstruction." *Geochimica et Cosmochimica Acta* 68 (2004): 61–72.

Johns, Adrian. "Miscellaneous Methods: Authors, Societies and Journals in Early Modern England." *British Journal for the History of Science* 33 (2000): 159–186.

Johnston, Wendy K., Peter J. Unrau, Michael S. Lawrence, Margaret E. Glasner, and David P. Bartel. "RNA-Catalyzed Polymerization: Accurate and General RNA-Templated Primer Extension." *Science* 292 (2001): 1319–1325.

Judson, Horace Freeland. *The Eighth Day of Creation: Makers of the Revolution in Biology*. 25th anniversary expanded ed. Plainview, N.Y.: Cold Spring Harbor Press, 1996.

Kearey, Philip, Keith A. Klepeis, and Frederick J. Vine. *Global Tectonics*. 3rd ed. Hoboken, N.J.: Wiley-Blackwell, 2009.

Kimura, Motoo. "A Simple Method for Estimating Evolutionary Rates of Base Substitutions through Comparative Studies of Nucleotide Sequences." *Journal of Molecular Evolution* 16 (1980): 111–120.

Kirschner, Marc W., and John C. Gerhart. *The Plausibility of Life: Resolving Darwin's Dilemma*. New Haven, Conn.: Yale University Press, 2005.

Klein, Nicole, and Martin Sander. "Ontogenetic Stages in the Long Bone Histology of Sauropod Dinosaurs." *Paleobiology* 34 (2008): 247–263.

Klein, Nicole, Kristian Remes, Carole T. Gee, P. Martin Sander, Oliver Wings, Andrá Borbély, Thomas Breuer, et al. *Biology of the Sauropod Dinosaurs: Understanding the Life of Giants*. Bloomington: Indiana University Press, 2011.

Klug, Aaron. "Rosalind Franklin and the Discovery of the Structure of DNA." *Nature* 219 (1968): 808–810.

Knoll, Andrew H. *Life on a Young Planet: The First Three Billion Years of Evolution on Earth*. Princeton, N.J.: Princeton University Press, 2003.

Koch, Paul L., Noreen Tuross, and Marilyn L. Fogel. "The Effects of Sample Treatment and Diagenesis on the Isotopic Integrity of Carbonate in Biogenic Hydroxylapatite." *Journal of Archaeological Science* 24 (1997): 417–429.

Koga, Shogo, David S. Williams, Adam W. Perriman, and Stephen Mann. "Peptide-Nucleotide Microdroplets as a Step towards a Membrane-Proof Protocell Module." *Nature Chemistry* 3 (2011): 720–724.

Kohn, Matthew J. "Predicting Animal δ18o: Accounting for Diet and Physiological Adaptation." *Geochimica et Cosmochimica Acta* 60 (1996): 4811–4829.

Kopal, Zdenek. "Ole Rømer." In *Dictionary of Scientific Biography*, edited by Charles Coulston Gillispie, 11:525–527. New York: Scribner, 1970.

Lalueza-Fox, Carles, Antonio Rosas, Almudena Estalrrich, Elena Gigli, Paula F. Campos, Antonio Garcia-Tabernero, Samuel Garcia Vargas, et al. "Genetic Evidence for Patrilocal Mating Behavior among Neandertal Groups." *Proceedings of the National Academy of Science* 108 (2011): 250–253.

Lane, Nick. *Power, Sex, Suicide: Mitochondria and the Meaning of Life.* Oxford: Oxford University Press, 2005.

Lawrence, Michael S., and David P. Bartel. "Processivity of Ribozyme-Catalyzed RNA Polymerization." *Biochemistry* 42 (2003): 8748–8755.

Lehman, Thomas M., and Holly N. Woodward. "Modeling Growth Rates for Sauropod Dinosaurs." *Paleobiology* 34 (2008): 264–281.

Levene, Phoebus A., and Lawrence W. Bass. *Nucleic Acids.* New York: Chemical Catalog, 1931.

Lewis, Ricki. "DNA Fingerprints: Witness for the Prosecution." *Discover*, June 1988, 44–52.

Li, Mei, David C. Green, J. L. Ross Anderson, Bernard P. Binks, and Stephen Mann. "*In Vitro* Gene Expression and Enzyme Catalysis in Bioinorganic Protocells." *Chemical Science* 2 (2011): 1739–1745.

Lingham-Soliar, Theagarten, and Joanna Glab. "Dehydration: A Mechanism for the Preservation of Fine Detail in Fossilized Soft Tissue of Ancient Terrestrial Animals." *Palaeogeography, Palaeoclimatology, Palaeoecology* 291 (2010): 481–487.

Livio, Mario. *Brilliant Blunders: From Darwin to Einstein—Colossal Mistakes by Great Scientists That Changed Our Understanding of Life and the Universe.* New York: Simon and Schuster, 2013.

Lu, Junchang, David M. Unwin, Denis Charles Deeming, Xingsheng Jin, Yongqing Liu, and Qiang Ji. "An Egg-Adult Association, Gender, and Reproduction in Pterosaurs." *Nature* 331 (2011): 321–324.

Lyell, Charles. *Principles of Geology.* 3 vols. Chicago: University of Chicago Press, 1990–1991.

Mackay, Andrew. "The Method of Finding the Longitude of a Place, by the Eclipses of the Satellites of Jupiter." In *The Theory and Practice of Finding the Longitude at Sea or Land; to Which Are Added, Various Methods of Determining the Latitude of a Place and Variation of the Compass; with New Tables*, 193–198. London: Sewell, Cornhill, P. Elmsly, Strand and J. Evans, 1793.

Maddox, Brenda. "The Double Helix and the 'Wronged Heroine.'" *Nature* 421 (2003): 407–408.

Malthus, Thomas R. *An Essay on the Principle of Population; or, A View of Its Past and Present Effects on Human Happiness.* 9th ed. London: Reeves and Turner, 1888.

——. *An Essay on the Principle of Population; or, A View of Its Past and Present Effects on Human Happiness: with an Inquiry into Our Prospects respecting the Future Removal or Mitigation of the Evils Which It Occasions.* Vols. 1, 2. Cambridge: Cambridge University Press for the Royal Economic Society, 1989.

——. *The Works of Thomas Malthus.* Edited by E. A. Wrigley and David Souden. Vols. 1, 2. London: W. Pickering, 1986.

Manning, Phillip L., Peter M. Morris, Adam McMahon, Emrys Jones, Andy Gize, Joe II. S. Macquaker, George Wolff, et al. "Mineralized Soft-Tissue Structure and Chemistry in a Mummified Hadrosaur from the Hell Creek Formation, North Dakota (USA)." *Proceedings of the Royal Society B: Biological Sciences* 276 (2009): 3429–3437.

Martin, Anthony J. *Dinosaurs without Bones: Dinosaur Lives Revealed by Their Trace Fossils.* New York: Pegasus Books, 2014.

Martins, Zita, Conel M. O'D. Alexander, Graznya E. Orzechowska, Marilyn L. Fogel, and Pascale Ehrenfreund. "Indigeneous Amino Acids in Primitive CR Meteorites." *Meteoritics and Planetary Science* 42 (2007): 2125–2136.

Marvin, Ursula B. *Continental Drift: The Evolution of a Concept.* Washington, D.C.: Smithsonian Institution Press, 1973.

Marvin, Ursula. "Meteorites in History: An Overview from the Renaissance to the 20th Century." In *The History of Meteoritics and Key Meteorite Collections: Fireballs, Falls and Finds*, edited by G. J. H. McCall, A. J. Bowden, and R. J. Howarth, 15–71. London: Geological Society, 2006.

Maskelyne, Nevil. "Directions for Observing the Beginning and Ending of an Eclipse of the Moon, and an Immersion or Emersion of the Satellites of Jupiter." In *The British Mariner's Guide: Containing Complete and Easy Instructions for the Discovery of Longitude at Sea and Land, within a Degree, by Observations of the Distance of the Moon from the Sun and Stars, Taken with Hadley's Quadrant. . . .* , 86–91. London: Printed for the author, 1763.

Maxmen, Amy. "Evolution: A Can of Worms." *Nature* 470 (2011): 161–162.

Mayor, Adrienne. *The First Fossil Hunters: Paleontology in Greek and Roman Times.* Princeton, N.J.: Princeton University Press, 2000.

McBride, Heidi M., Margaret Neuspiel, and Sylwia Wasiak. "Mitochondria: More Than Just a Powerhouse." *Current Biology* 16 (2005): R551–R560.

McElhinny, Michael W. *Paleomagnetism: Continents and Oceans.* San Diego, Calif.: Academic Press, 2000.

McNab, Brian K. "Resources and Energetics Determined Dinosaur Maximal Size." *Proceedings of the National Academies of Sciences* 106 (2009): 12184–12188.

Mehra, Jagdish. *Einstein, Hilbert, and the Theory of Gravitation: Historical Origins of General Relativity Theory.* Dordrecht: Reidel, 1974.

Mendel, Gregor. "Experiments in Plant Hybridization." *Journal of the Royal Horticultural Society* 26 (1901): 1–32.

Miller, Stanley Lloyd. "A Production of Amino Acids under Possible Primitive Earth Conditions." *Science* 117 (1953): 528–529.

——. "Production of Some Organic Compounds under Possible Primitive Earth Conditions." *Journal of the American Chemical Society* 77 (1955): 2351–2361.

——. "Which Organic Compounds Could Have Occurred on the Prebiotic Earth?" *Cold Spring Harbor Symposia on Quantitative Biology* 52 (1987): 17–27.

Mills, Donald R., Roger L. Peterson, and Sol Spiegelman. "An Extracellular Darwinian Experiment with a Self-Duplicating Nucleic Acid Molecule." *Proceedings of the National Academy of Science* 58 (1967): 217–224.

Morgan, Thomas Hunt. *The Mechanism of Mendelian Heredity.* Rev. ed. New York: H. Holt, 1923.

Mukherjee, Siddhartha. *The Gene: An Intimate History.* New York: Scribner, 2016.

Muñoz Caro, Guillermo Manuel, Uwe J. Meierhenrich, Winfried A. Schutte, Bernard Barbier, Angel Arcones Segovia, Helmut Rosenbauer, Wolfram H.-P. Thiemann, André Brack, and Jerome Mayo Greenberg. "Amino Acids from Ultraviolet Irradiation of Interstellar Ice Analogues." *Nature* 416 (2002): 403–406.

Nersessian, Nancy J. *Creating Scientific Concepts.* Cambridge, Mass.: MIT Press, 2008.

Neugebauer, Otto. "The Egyptian 'Decans.'" *Vistas in Astronomy* 1 (1955): 47–51.

——. *The Exact Sciences in Antiquity.* 2nd ed. Providence, R.I.: Brown University Press, 1957.

——. *A History of Ancient Mathematical Astronomy.* New York: Springer-Verlag, 1975.

Nikaido, Masato, Alejandro P. Rooney, and Norihiro Okada. "Phylogenetic Relationships among Cetartiodactyls Based on Insertions of Short and Long Interspersed Elements: Hippopotamuses Are the Closest Extant Relatives of Whales." *Proceedings of the National Academies of Sciences of the United States* 96 (1999): 10261–10266.

Norell, Mark A. "Tree-Based Approaches to Understanding History: Comments on Ranks, Rules, and the Quality of the Fossil Record." *American Journal of Science* 293-A (1993): 407–417.

North, John David. *Cosmos: An Illustrated History of Astronomy and Cosmology.* Chicago: University of Chicago Press, 2008.

Novelline, Robert A. *Squire's Fundamentals of Radiology.* 6th ed. Cambridge, Mass.: Harvard University Press, 2004.

Novistki, Edward. "On Fisher's Criticism of Mendel's Results with the Garden Pea." *Genetics* 166 (2004): 1133–1136.

Numbers, Ronald L., and Kostas Kampourakis, eds. *Newton's Apple and Other Myths about Science*. Cambridge, Mass.: Harvard University Press, 2015.

Olby, Robert C. "DNA before Watson-Crick." *Nature* 248 (1974): 782–785.

———. *The Path to the Double Helix: The Discovery of DNA*. New York: Dover Publications, 1994.

Oldroyd, David Roger. *Thinking about the Earth: A History of Ideas in Geology*. Cambridge, Mass.: Harvard University Press, 1996.

O'Leary, Maureen A., and Mark D. Uhen. "The Time of Origin of Whales and the Role of Behavioral Changes in the Terrestrial-Aquatic Transition." *Paleobiology* 25 (1999): 534–556.

Orgel, Leslie E. "The Origin of Life on Earth." *Scientific American* 271 (1994): 76–83.

Pääbo, Svante. *Neanderthal Man: In Search of Lost Genomes*. New York: Basic Books, 2014.

Pais, Abraham. *"Subtle Is the Lord": The Science and the Life of Albert Einstein*. Oxford: Clarendon Press of Oxford University Press, 1982.

Pauling, Linus C. *Molecular Architecture and the Processes of Life*. Nottingham, U.K.: Sir Jesse Boot Foundation, 1948.

———. "Molecular Basis of Biological Specificity." *Nature* 248 (1974): 769–771.

Perry, John. "On the Age of the Earth." *Nature* 51 (1895): 224–227.

Phinney, Robert A., ed. *The History of the Earth's Crust: A Symposium*. Princeton, N.J.: Princeton University Press, 1968.

Pierazzo, Elisabetta, and Christopher F. Chyba. "Amino Acid Survival in Large Cometary Impacts." *Meteoritics and Planetary Science* 34 (1999): 909–918.

Pontzer, Herman, Vivian Allen, and John R. Hutchinson. "Biomechanics of Running Indicates Endothermy in Bipedal Dinosaurs." *PLoS ONE* 4, no. 11: e7783.

Poole, Anthony M., and Derek T. Logan. "Modern mRNA Proofreading and Repair: Clues That the Last Universal Common Ancestor Possessed an RNA Genome?" *Molecular Biology and Evolution* 22 (2005): 1444–1455.

Portugal, Franklin H., and Jack S. Cohen. *A Century of DNA: The History of the Discovery of the Structure and Function of the Genetic Substance*. Cambridge, Mass.: MIT Press, 1977.

Pyron, R. Alexander. "A Likelihood Method for Assessing Molecular Divergence Time Estimates and the Placement of Fossil Calibrations." *Systematic Biology* 59 (2010): 185–194.

Quammen, David. *The Tangled Tree: A Radical New History of Life*. New York: Simon and Schuster, 2018.

Reisz, Robert R., Dianne Scott, Hans-Dieter Sues, David C. Evans, and Michael A. Raath. "Embryos of an Early Jurassic Prosauropod Dinosaur and Their Evolutionary Significance." *Science* 909 (2005): 761–764.

Reston, James, Jr. *Galileo: A Life*. London: Cassell, 1994.

Roche, John J. "Harriot, Galileo, and Jupiter's Satellites." *Archives Internationales d'Histoire des Sciences* 32 (1982): 9–51.

Rogers, Everett M. *Diffusion of Innovations*. 5th ed. New York: Free Press, 2003.

Romer, M., and I. Bernard Cohen. "Roemer and the First Determination of the Velocity of Light." *Isis* 31 (1940): 327–379.

Rowan-Robinson, Michael. *The Cosmological Distance Ladder: Distance and Time in the Universe*. New York: W. H. Freeman, 1984.

Rudwick, Martin J. S. *Bursting the Limits of Time: The Reconstruction of Geohistory in the Age of Revolution*. Chicago: University of Chicago Press, 2005.

——. *Georges Cuvier, Fossil Bones, and Geological Catastrophes: New Translations and Interpretations of the Primary Texts*. Chicago: University of Chicago Press, 1997.

Rutherford, Ernest. "Radioactive Change." *Philosophical Magazine* 6 (1903): 576–591.

Rutledge, James. "De la chaise marine." In *Nouvelle théorie astronomique ou servir à la determination des longitudes*, 98–103. Paris: Chez Volland, 1788.

Ruxton, Graeme D., and David M. Wilkinson. "The Energetics of Low Browsing in Sauropods." *Biology Letters* 7 (2011): 779–781.

Rømer, Ole. "Demonstration touchant le mouvement de la lumière trouvé par M. Roemer de l'Academie Royale des Sciences." *Journal des Scavans* 269 (1676): 233–236.

San Antonio, James D., Mary H. Schweitzer, Shane T. Jensen, Raghu Kalluri, Michael Buckley, and Joseph P. R. O. Orgel. "Dinosaur Peptides Suggest Mechanisms of Protein Survival." *PLoS ONE* 6 (2011): e20381.

Sander, P. Martin, Andreas Christian, Marcus Clauss, Regina Fechner, Carole T. Gee, Eva-Maria Griebler, Hanns-Christian Gunga, et al. "Biology of the Sauropod Dinosaurs: The Evolution of Gigantism." *Biological Reviews* 86 (2011): 117–155.

Sander, P. Martin, and Marcus Clauss. "Sauropod Gigantism." *Science* 322 (2008): 200–201.

Schechner, David M., and David P. Bartel. "The Structural Basis of RNA-Catalyzed RNA Polymerization." *Nature Structural and Molecular Biology* 18 (2011): 1036–1042.

Scheffler, Immo E. *Mitochondria*. 2nd ed. Hoboken, N.J.: Wiley-Liss, 2008.

Scherer, Stewart. *A Short Guide to the Human Genome*. Cold Spring Harbor, N.Y.: Cold Spring Harbor Press, 2008.

Schindler, Samuel. "Model, Theory and Evidence in the Discovery of the DNA Structure." *British Journal for the Philosophy of Science* 59 (2008): 619–658.

Schopf, J. William. *Cradle of Life: Discovery of Earth's Earliest Fossils*. Princeton, N.J.: Princeton University Press, 1999.

——. "Deep Divisions in the Tree of Life: What Does the Fossil Record Reveal?" *Biological Bulletin* 196 (1999): 351–355.

——, ed. *Major Events in the History of Life*. Boston: Jones and Bartlett, 1992.

——. "Microfossils of the Early Archean Apex Chert: New Evidence of the Antiquity of Life." *Science* 260 (1993): 640–646.

Schultes, Erik A., and David P. Bartel. "One Sequence, Two Ribozymes: Implications for the Emergence of New Ribozyme Folds." *Science* 289 (2000): 448–452.

Schultz, Peter G., and Richard A. Lerner. "From Molecular Diversity to Catalysis: Lessons from the Immune System." *Science* (1995): 1835–1842.

Schwartz, James. *In Pursuit of the Gene: From Darwin to DNA.* Cambridge, Mass.: Harvard University Press, 2008.

Schweitzer, Mary Higby, Jennifer L. Wittmeyer, and John R. Horner. "Soft Tissue and Cellular Preservation in Vertebrate Skeletal Elements from the Cretaceous to the Present." *Proceedings of the Royal Society B: Biological Sciences* 274 (2007): 183–197.

Seymour, Roger S., Sarah L. Smith, Craig R. White, Donald M. Henderson, and Daniela Schwarz-Wings. "Blood Flow to Long Bones Indicates Activity Metabolism in Mammals, Reptiles and Dinosaurs." *Proceedings of the Royal Society B: Biological Sciences* 279 (2012): 451–456.

Shapley, Harlow, and Heber Doust Curtis. "The Scale of the Universe." *Bulletin of the National Research Council* 11 (1921): 171–217.

Sharp, Phillip A. "Split Genes and RNA Splicing." *Cell* 77 (1994): 805–815.

Shubin, Neil. *Your Inner Fish: A Journey into the 3.5-Billion-Year History of the Human Body.* New York: Pantheon, 2008.

Silliman, Benjamin. "Memoir on the Origin and Composition of the Meteoric Stones Which Fell from the Atmosphere." *Transactions of the American Philosophical Society* 6 (1809): 235–245.

Simpson, George Gaylord. "Mammals and the Nature of Continents." *American Journal of Science* 241 (1943): 1–31.

Sitter, Willem de. "Jupiter's Galilean Satellites (George Darwin Lecture)." *Monthly Notices of the Royal Astronomical Society* 91 (1931): 706–738.

Sivin, Nathan. *Granting the Seasons: The Chinese Astronomical Reform of 1280, with a Study of Its Many Dimensions and a Translation of Its Records.* New York: Springer, 2009.

Smith, Vincent S., Tom Ford, Kevin P. Johnson, Paul C. D. Johnson, Kazunori Yoshizawa, and Jessica E. Light. "Multiple Lineages of Lice Pass through the K-Pg Boundary." *Biology Letters* 7 (2005): 782–785.

Solow, Andrew R., and Michael J. Benton. "On the Flux Ratio Method and the Number of Valid Species Names." *Paleobiology* 36 (2010): 516–518.

Spicer, Robert A., and Alexei B. Herman. "The Late Cretaceous Environment of the Arctic: A Quantitative Reassessment Based on Plant Fossils." *Palaeogeology, Palaeoclimatology, Palaeoecology* 295 (2010): 423–442.

Spotila, James R., Michael P. O'Connor, Peter Dodson, and Frank V. Paladino. "Hot and Cold Running Dinosaurs: Body Size, Metabolism and Migration." *Modern Geology* 16 (1991): 203–227.

Stachel, John. *Einstein from "B" to "Z."* Boston: Birkhäuser, 2002.

Stent, Gunther S. "Molecular Biology and Metaphysics." *Nature* 248 (1974): 779–781.

Stone, Marcia. "Life Redesigned to Suit the Engineering Crowd." *Microbe* 1 (2006): 566–570.

Struve, Otto. "First Determinations of Stellar Parallax—I." *Sky and Telescope* 11 (1956): 9–11.

———. "First Determinations of Stellar Parallax—II." *Sky and Telescope* 11 (1956): 69–72.

Tarver, James E., Philip C. J. Donoghue, and Michael J. Benton. "Is Evolutionary History Repeatedly Rewritten in Light of New Fossil Discoveries?" *Proceedings of the Royal Society B: Biological Sciences* 278 (2011): 599–604.

Taton, René, and Curtis Wilson. *Planetary Astronomy from the Renaissance to the Rise of Astrophysics.* New York: Cambridge University Press, 1989.

Taylor, Eva Germaine Rimington. *Mathematical Practitioners of Hanoverian England, 1714–1840.* London: Cambridge University Press, 1966.

Theobold, Douglas L. "A Formal Test of the Theory of Universal Formal Ancestry." *Nature* 465 (2010): 219–222.

Thomas, Roger D. K., and Everett C. Olson, eds. *A Cold Look at the Warm-Blooded Dinosaurs.* Boulder, Colo.: Westview, 1980.

Thomson, William. "On the Secular Cooling of the Earth." *London, Edinburgh and Dublin Philosophical Magazine and Journal of Science* 25 (1863): 1–14.

Tian, Feng, Owen B. Toon, Alexander A. Pavlov, and Hans De Sterck. "A Hydrogen-Rich Early Earth Atmosphere." *Science* 308 (2005): 1014–1017.

Toomer, Gerald J. "Addenda and Corrigenda." In *Ptolemy's Almagest,* xi–xiv. London: Duckworth, 1984.

Trifonov, Edward N. "The Triplet Code from First Principles." *Journal of Biomolecular Structure and Dynamics* 22 (2004): 1–11.

Upchurch, Paul. "The Evolutionary History of Sauropod Dinosaurs." *Philosophical Transactions: Biological Sciences* 349 (1995): 365–390.

Vaiden, Robert. *Plate Tectonics: Mysteries Solved!* Illinois State Geological Survey, Geobit 10, 2004.

Valleriani, Matteo. *Galileo Engineer.* Dordrecht: Springer, 2010.

Van Helden, Albert. *Measuring the Universe: Cosmic Dimensions from Aristarchus to Halley.* Chicago: University of Chicago Press, 1985.

Vanpaemel, Geert. "Science Disdained: Galileo and the Problem of Longitude." In *Italian Scientists in the Low Countries in the XVIIth and XVIIIth Centuries,* edited by C. S. Maffeoli and L. C. Palm, 111–129. Amsterdam: Rodopi, 1989.

Vinci, Leonardo da. *The Codex Leicester: Notebook of a Genius.* Sydney, N.S.W.: Powerhouse, 2000.

———. *The Notebooks of Leonardo da Vinci.* Compiled and edited from the original manuscripts by Jean Paul Richter. 2 vols. New York: Dover, 1970.

Vine, Frederick J. "Spreading of the Ocean Floor: New Evidence." *Science* 154 (1966): 1405–1415.

Vine, Frederick J., and Drummond H. Matthews. "Magnetic Anomalies over Ocean Ridges." *Nature* 199 (1963): 947–949.

Vischer, Ernst, and Erwin Chargaff. "The Separation and Characterization of Purines in Minute Amounts of Nucleic Acid Hydrolysates." *Journal of Biological Chemistry* 168 (1947): 781.

Wacey, David, Matt R. Kilburn, Martin Saunders, John Cliff, and Martin D. Brasier. "Microfossils of Sulphur-Metabolizing Cells in 3.4-Billion-Year-Old Rocks of Western Australia." *Nature Geoscience* 4 (2011): 698–702.

Wagner, Robert P. "Genetics and Phenogenetics of Mitochondria." *Science* 163 (1969): 1026–1031.

Wallace, Alfred Russell. "On the Tendency of Varieties to Depart from the Original Type." *Proceedings of the Linnaean Society of London* 3 (1858): 53–62.

Wang, James C. "Interaction between DNA and an Escherichia coli Protein Omega." *Journal of Molecular Biology* 55 (1971): 523–533.

Wang, Lei, Jianming Xie, and Peter G. Schultz. "Expanding the Genetic Code." *Annual Review of Biophysics and Biomolecular Structure* 35 (2006): 225–249.

Wang, Xiaolin, and Zhonghe Zhou. "Pterosaur Embryo from the Early Cretaceous." *Nature* 429 (2004): 621.

Watson, James Dewey. *DNA: The Secret of Life.* New York: Alfred A. Knopf, 2003.

———. *Molecular Biology of the Gene.* 6th ed. Cold Spring Harbor, N.Y.: Cold Spring Harbor Laboratory Press, 2008.

Watson, James Dewey, and Francis Harry Compton Crick. "Genetical Implications of the Structure of Deoxyribonucleic Acid." *Nature* 171 (1953): 964–967.

———. "Molecular Structure of Nucleic Acids: A Structure for Deoxyribose Nucleic Acid." *Nature* 171 (1953): 737–738.

Wegener, Alfred. "Entstehung der Kontinente." *Geologische Rundschau* 3 (1912): 276–292.

———. *The Origin of Continents and Oceans.* Translated from the 4th rev. German ed. by John Biram. New York: Dover, [1966].

Weishampel, David B., Peter Dodson, and Halszka Osmólska, eds. *Dinosauria.* Berkeley: University of California Press, 1990.

———. *Dinosauria.* 2nd ed. Berkeley: University of California Press, 2004.

Westall, Frances, Maarten J. de Wit, Jesse Dann, Sjerry van der Gaast, Cornel E. J. de Ronde, and Dane Gerneke. "Early Archean Fossil Bacteria and Biofilms in Hydrothermally-Influenced Sediments from the Barberton Greenstone Belt, South Africa." *Precambrian Research* 106 (2001): 93–116.

Wilford, John Noble. "Giants Who Scarfed Down Fast Food." *New York Times,* April 11, 2011.

Wilkins, Maurice Hugh Frederick, Alec Rawson Stokes, and Herbert Rees Wilson. "Molecular Structure of Nucleic Acids: Molecular Structure of Deoxypentose Nucleic Acids." *Nature* 171 (1953): 738–740.

Williams, Robert Joseph Paton, and João J. R. Frausto da Silva. *Bringing Chemistry to Life: From Matter to Man.* Oxford: Oxford University Press, 1999.

Wiltschi, Birgit, and Nediljko Budisa. "Natural History and Experimental Evolution of the Genetic Code." *Applied Microbiology and Biotechnology* 74 (2007): 739–753.

Woese, Carl R. "Bacterial Evolution." *Microbiological Reviews* 51 (1987): 221–271.

Wong, Jeffrey Tze-Fei. "Coevolution of the Genetic Code at Age Thirty." *BioEssays* 27 (2005): 416–425.

Xu, Xing, Zhi-Lu Tang, and Xiao-Lin Wang. "A Therizinosaurid Dinosaur with Integumentary Structures from China." *Nature* 399 (1999): 350–354.

Xu, Xing, Kebai Wang, Ke Zhang, Qingyu Ma, Lida Xing, Corwin Sullivan, Dongyu Hu, Shuqing Cheng, and Shuo Wang. "A Gigantic Feathered Dinosaur from the Lower Cretaceous of China." *Nature* 484 (2012): 92–95.

Xu, Xing, and Guo Yu. "The Origin and Early Evolution of Feathers: Insights from Recent Paleontological and Neontological Data." *Vertebrata PalAsiatica* 47 (2009): 311–329.

Yaffe, Michael P. "The Machinery of Mitochondrial Inheritance and Behavior." *Science* 283 (1999): 1493–1497.

INDEX

Page numbers in italics indicate figures. Pages numbered in roman type may also contain figures.

earthquakes: amplitude of energy released, 142–43; hydraulic fracturing (fracking) and, 144; hydraulic fracturing (fracking) and, 145–47; locations of, 139–41, 163; seismic waves, 133, 135–40; theories of, 133. *See also* seismometers

earth's surface, erosion and volcanism, 148

eclipses: Io (Jupiter), 49–52; solar and lunar, 16, 74, 127–28

Einstein, Albert: expansion of the universe, 115; famous formula, 117; general theory of relativity, 72–73, 90–91, 108; motion of water particles, 172; nature's understandability, 7; Principle of Equivalence and, 59; special theory of relativity, 70–71, 254; speed of light and, 94; supernova data and, 119; testable predictions of, 74

electrical force, 120–21

Eltanin (research ship), 159, 162

Encke, Johann, 65

enzymes, 253, 307

Eratosthenes, 128–31

Erickson, Gregory, 211

Escherichia coli (E. coli), 276, 278, 302–4, 312

European Space Agency, 83, 108

evolution: cause of, 238; critique of, 245; phylogeny, 289, 294; plague genome, 310; tests of, 302–4. *See also* bears; whales

evolutionary biology, 260

Ewen, Harold Irving "Doc," 321

extinction, 198, 202

extremophiles, 287–88, 300

false claims, 168–69

Faraday, Michael, 70

Fermi, Enrico, 328

Fermi paradox, 328

Field, George, 106

The First Three Minutes (Weinberg), 100

Fischer, Al, 220

Fisher, Osmond, 150

Fisher, R. A., 249

Ford, Kent, 111–12

fossils: anatomy of, 202; definition of, 197; embryology of, 294; preservation of, 197–200; study of, 201–2; trilobites, 198. *See also* bears; whales

Foulke, William, 205

Frail, Dale, 325

Franklin, Benjamin, 62, 174

Franklin, Rosalind, 266, 268–70, 273–75

Fraunhofer, Joseph von, 91

French Academy of Sciences, 52

Friedmann, Alexander, 91, 95

Fuller, Buckminster, 322

Furberg, Sven, 267–68

Gaia (spacecraft), 83–84, 88

galaxy: antigravity and, 118; distance calculation, 76, 97–99, 111–13, 116; gravitational binding of, 110–11; gravitational force, 114; hydrogen atoms in, 111; mass distribution, 112–13; Milky Way, 76, 113; one or many, 89–91, 94; pancake shape, 89; peculiar velocities of, 96; rotation curves, 112–13; solar system's location, 90; spectra of, 91–92, 94, 110. *See also* Cepheid variables; dark matter

Galileo, 43–48, 52, 59, 66

Galle, Johann, 64–66

Gamow, George, 100, 277

Geiger, Hans, 175

gene editing, 312–13

general theory of relativity: cosmological constant, 90, 119; gravitation and, 73, 108; model of nature, 179; principle of equivalence and, 72

geological time, 221

geology: age of earth, 188, 191; catastrophism, 202; Chicxulub impact crater, 226–28, 231, 233; contractionists, 149; Hell Creek formation, 232; KT layer, 220–21, 223; permanentists, 149; prestige of, 148; stratigraphy, 202; uniformitarians, 149

geophysics and geophysicists, 139, 153–54

geopoetry, 159

Gesner, Conrad, 201–2

Gilbert, William, 39